DATE			

AMERICAN ENGINEERS
OF THE NINETEENTH CENTURY

GARLAND REFERENCE LIBRARY
OF SOCIAL SCIENCE
(VOL. 53)

AMERICAN ENGINEERS
OF THE NINETEENTH CENTURY
A Biographical Index

Christine Roysdon
Linda A. Khatri

GARLAND PUBLISHING, INC. • NEW YORK & LONDON
1978

Library of Congress Cataloging in Publication Data

Roysdon, Christy.
    American engineers of the nineteenth century.

    (Garland reference library of social science; v. 53)
    1. Engineers—United States—Biography—Indexes.
I. Khatri, Linda A., joint author. II. Title.
Z5851.R7 [TA139]   016.62'00092'2   77-83371
ISBN 0-8240-9827-7

To Rosemary

CONTENTS

PREFACE

During the late nineteenth century, brief biographies of several thousand American engineers and technologists appeared in the technical and trade press. The lack of indexes to this material has presented a barrier to its effective use by scholars. Cumulative indexes exist for each of the journals of the major engineering societies, but there is no consolidated index that covers all, as the Engineering Index now does for current technical literature. Trade journals pose an even greater problem, as only a few titles have cumulative indexes, from which references to biographies must be painstakingly plucked. Most have volume indexes, but some have none at all. The result is that the biographer or historian desiring to consult these sources has been obliged to engage in tedious page-by-page searching.

Scholars who have braved the difficulties of recovering biographies from these journals have no doubt found the work rewarding. The trade journals chronicle the work and lives of individuals whose achievements are doubtless significant, even well known, yet who remain anonymous or obscure. Nineteenth-century cyclopedias of biography, while helpful, are of course incomplete. Later comprehensive works, such as the Dictionary of American Biography, tend to include only the most influential and wealthy technologists. Society journals profile only their members. Thus, it is hoped that our index will not only help solve problems of access, but also open opportunities for scholars to study and assess the work of hitherto little known, potentially important figures.

Of the vast array of technical periodicals that were founded in the United States during the nineteenth century, candidates for indexing were narrowed to those that enjoyed fairly wide, national circulation, had substantial nineteenth-century runs, and featured regular obituary or biographical columns. Among these, we sought a selection of periodicals that would provide full coverage of the rapidly diversifying engineering profession and of all major areas of technological activity.

Inclusion in the index was limited to engineers and technologists who died during or before 1900. Admittedly, this date is arbitrary and results in exclusion of some important nineteenth-century figures who died within the subsequent few decades. The choice was based mostly on the desire to keep the size of the file manageable, as adding another five or ten years to the period covered might have doubled it.

As a result, those persons indexed were largely engaged in activities that characterized mid- and late-nineteenth-century technology, especially the development of railroads and canals. The process of industrialization, particularly the rise of iron and steel, is also strongly represented by the large number of manufacturers. Electrical engineering, still in its infancy, is only seminally represented; chemical engineering was virtually nonexistent.

Care was taken to select only those individuals whose careers were of a technical nature. Engineers and inventors of practical devices were automatically included as were industrialists who made significant contributions to the manufacturing process. Several important technical journalists were indexed on the basis of their close relationship with the technology of the period. An effort was made to eliminate those whose technical contributions were negligible, or of an ephemeral nature, such as capitalists, financiers, telegraphers, inventors of useless gadgets, and drivers of locomotives and canal boats.

The length, quality, and accuracy of biographical material vary greatly from journal to journal. Most of the biographies indexed are obituary notices, but there are also sketches and profiles. The choice biographies—the most complete and accurate—derive from the society journals. The most comprehensive treatment, numerically speaking, of both eminent and obscure engineers is provided by *Engineering Record*; similar coverage is achieved by *Iron Age* for iron manufacturers. Unfortunately, obituaries in these and some similar sources tend to be very short. Another caveat in utilizing trade journal biographical fragments is that periodicals tended to borrow from one another, so that similar or identical biographies of an individual may appear in several sources. As might be expected, there is a long list of citations for significant figures, such as Alexander Holley and John Ericsson. To a degree, the number of biographical notices an individual received is an index of his importance. But researchers will note that some illustrious persons, such as Samuel F.B. Morse, who died at a time when few periodicals had yet begun publishing obituaries, have very few citations. The bulk of the obituary notices date from the period 1880 to 1900.

Each entry is followed by the abbreviated titles for the sources of information, along with the year, volume number, and inclusive pages. For noncontinuously paginated volumes, the date of the issue precedes the year. An asterisk (*) is used to indicate an accompanying portrait. The supplementary sources follow a similar format, with the exclusion of the year. The one exception is the *New York Times*, in which case the date of the issue is followed by page and column numbers.

In addition to the sources indexed here, biographical information can be found in dictionaries such as the *National Cyclopedia of American Biography* and *Lamb's Dictionary of American Biography*. The Committee on History and Heritage of the American Society of Civil Engineers has published *A Biographical Dictionary of American Civil Engineers* (1972), with fairly lengthy information on the lives of 170 civil engineers born prior to the Civil War. Mention is made in the prefatory remarks of projected volumes of a similar nature. Eugene Ferguson's *Bibliography of the History of Technology* (1968) contains an excellent section on sources for bibliographical information, including guides to dictionaries, bibliographies, and full-length biographies. Not to be overlooked is the file of biographical data on American engineers maintained at the National Museum of History and Technology. For British engineers of the same period, S. Peter Bell has produced an index similar in nature to this one, drawing upon information contained in a number of the major British trade and society technical journals.

We wish to thank those who have helped to make the task somewhat more pleasant. Special thanks go to Professor Eugene Ferguson of the Hagley Museum for his consultation and encouragement, and to Mr. Robert Vogel of the Smithsonian Institution for his hospitality during our trip to Washington. The cooperation of the many libraries we visited, especially the Engineering Societies Library in New York and Skillman Library of Lafayette College in Easton, Pa., was greatly appreciated. To the many others who helped carry the endless, dusty volumes from their remote habitats, and who exhibited much patience and feigned pleasure in hearing the tales of the lives of nineteenth-century engineers, we are forever indebted.

Christine Roysdon
Linda A. Khatri
Lehigh University
Bethlehem, Pennsylvania
January, 1978

JOURNALS INDEXED

American Engineer and Railroad
Journal. 1832-1900; vols. 1-74.

Am Eng & RR J

American Institute of Electrical
Engineers. Transactions.
1884-1900; vols. 1-17.

AIEE Trans

American Institute of Mining
Engineers. Transactions.
1871-1900; vols. 1-30.

AIME Trans

American Iron and Steel Association.
Bulletin. 1866-1900; vols. 1-34.

AISA Bull

American Machinist.
1877-1900; vols. 1-23.

Am Mach

American Manufacturer and Iron
World. 1873-7, 1880-1900;
vols. 13-14, 16-20, 26-67.

Am Manuf

American Society of Civil
Engineers. Proceedings.
1873-1900; vols. 1-26.

ASCE Proc

American Society of Civil
Engineers. Transactions.
1867-1900; vols. 1-44.

ASCE Trans

American Society of Mechanical
Engineers. Transactions.
1880-1900; vols. 1-21.

ASME Trans

Appleton's Mechanics' Magazine
and Engineering Journal.
1851-3; vols. 1-3.

Appleton's
Mech Mag

Association of Engineering
   Societies. Journal.
   1881-1900; vols. 1-25.

Assoc Eng Soc J

Cassier's Engineering Monthly.
   1891-1900; vols. 1-17.

Cassier's

Electrical Engineer.
   1882-99; vols. 1-27.

Elec Eng

Electrical World.
   1883-1900; vols. 1-36.

Elec World

Engineer.
   1881-1900; vols. 1-37.

Eng

Engineering and Mining Journal.
   1886-1900; vols. 1-70.

Eng & Min J

Engineering News.
   1874-1900; vols. 1-44.

Eng News

Engineering Record.
   1888-1900; vols. 18-42.

Eng Rec

Franklin Institute. Journal.
   1826-1900; vols. 1-50.

Frank Inst J

Industrial World and Iron Worker.
   1880-1900; vols. 15-50.

Ind World

Iron Age.
   1873-1900; vols. 11, 13-66.

Iron Age

Machinery.
   1894-1900; vols. 1-6.

Mach

National Car and Locomotive Builder.
   1870-5, 1880-1, 1883-95;
   vols. 1-6, 11-12, 14-26.

Nat Car

Popular Science Monthly.
   1872-1900; vols. 1-57.

Pop Sci

Power.
   1884-91, 1893-1900;
   vols. 5-11, 13-20.

Power

Railroad Gazette.
   1870-1900; vols. 1-32.

RR Gaz

Sanitary Engineer.
   1877-88; vols. 1-17
   (became Engineering Record after
   vol. 17).

Sanit Eng

Scientific American.
   Old series: 1845-59; vols. 1-14.
   New series: 1859-1900; vols. 1-83.

Sci Am

Van Nostrand's Engineering Magazine.
   1869-86; vols. 1-35.

Van Nostrand's

Western Society of Engineers.
   Journal. 1896-1900; vols. 1-5.

West Soc Eng J

Western Society of Engineers.
   Proceedings. 1873-81; vols. 1-6.

West Soc Eng
Proc

SUPPLEMENTARY SOURCES

Appleton's Cyclopedia of American
   Biography.

Appleton's Cyc

Dictionary of American Biography.

DAB

New York Times.

NYT

Aaron, Charles H. (1833-93).   Miner, metallurgist, and
    author of works on assaying; inventor of process for
    treating silver ores.

    Eng & Min J (1891) 51:653*; (1893) 56:622

Abbott, Horace (1806-87).   Iron manufacturer, construc-
    tor of large rolling mills which produced armor plates
    for Monitor during Civil War.

    Am Eng & RR J (1887) 61:398
    Am Manuf (8/19/1887) 41:11
    RR Gaz (1887) 19:531
    Sanit Eng (1887) 16:298

    Appleton's Cyc 1:6
    DAB 1:21

Abbott, Job (1845-96).   Civil engineer, railroads and
    bridges.

    ASCE Proc (1896) 22:152-3
    ASCE Trans (1896) 36:538-9
    Eng Rec (1896) 34:253
    RR Gaz (1896) 28:610

Abend, Alexander E. (? -1895).   City engineer of East
    St. Louis.

    Eng Rec (1895) 32:327

Abendroth, John (1825?-1900).   Iron founder and boiler
    maker.

    Am Mach (1900) 23:753
    Eng & Min J (1900) 70:135
    Iron Age (7/26/1900) 66:15

Abendroth, William P. (1818?-98).   Iron manufacturer.

    Eng & Min J (1898) 65:168
    Eng Rec (1897-8) 37:205
    Iron Age (2/3/1898) 61:27

Abert, John James (1788-1862).   Military and topographi-
    cal engineer, coasts and harbors.

    ASCE Proc (1893) 19:88-92*

    Appleton's Cyc 1:8*

Ackenheil, Charles (1841-90). Chief engineer of Baltimore & New York Railroad.

Am Eng & RR J (1890) 64:331
ASCE Proc (1891) 17:137-8
RR Gaz (1890) 22:465

Adams, Albert S. (? -1899). Master mechanic of Boston & Worcester Railroad.

RR Gaz (1899) 31:817

Adams, Freeborn (1806?-91). Inventor of process for making seamless copper tubes.

Iron Age (1891) 48:1032

Adams, John M. (? -1892). Mining engineer.

Eng & Min J (1892) 54:132

Adams, Jonathan (1798?-1872). Civil engineer, canals in South and railroads in Northeast.

RR Gaz (1872) 4:413

Adams, Julius Walker (1812-99). Civil engineer, connected with several railroad, water works, and sewerage projects; projector of East River Suspension Bridge.

Am Eng & RR J (1900) 74:30
Eng & Min J (1899) 68:736
Eng News (1899) 42:400-1,403*
Eng Rec (1899) 40:665
Iron Age (12/21/1899) 64:21
RR Gaz (1899) 31:887*

Appleton's Cyc 1:28

Adams, Seth (1807-73). Inventor of power press; manufacturer of machinery and presses.

Sci Am (1873) 29:400

Appleton's Cyc 1:31

Adams, Wellington (1848?-98). Inventor and projector of Chicago and St. Louis elevated railroads.

Elec Eng (1898) 26:653

Addy, Matthew (1835-96). Manufacturer of pig iron and cast iron pipes.

AISA Bull (1896) 30:181
Am Manuf (1896) 59:195
Iron Age (1896) 58:273

Adelberg, Justus (1824?-69). Mining engineer.

Eng & Min J (1869) 7:376,406

NYT (6/7/1869) 5:5

Ahearn, John (1811?-95). Iron master in Baltimore.

AISA Bull (1895) 29:285
Eng & Min J (1895) 60:567
Iron Age (1895) 56:1223

Aiken, Walter (? -1893). Inventor of knitting machinery; president of Mt. Washington Railroad, N.H.

RR Gaz (1893) 25:936

Ainslie, George (1814-78). Civil engineer and mechanic.

ASCE Proc (1889) 15:41-2

Allaire, William Miller (1854-84). Civil engineer, involved with moving and erection of Egyptian obelisk in Central Park.

ASCE Proc (1893) 19:67

Allen, Henry Frederick (? -1894). Mechanical engineer.

Eng Rec (1893-4) 29:328

Allen, Horatio (1802-89). Civil engineer and inventor, influential in introducing first locomotives to U.S.; consultant for Brooklyn Bridge and Panama Railroad.

Am Eng & RR J (1890) 64:92
Am Mach (1/9/1890) 13:8
Am Manuf (1/10/1890) 46:13
ASCE Proc (1890) 16:180-4
ASME Trans (1890) 11:1156-81
Cassier's (1896) 10:402*,471-4
Eng (1889) 18:122*
Eng & Min J (1890) 49:32
Eng News (1890) 23:29-30*
Eng Rec (1889-90) 21:66
Iron Age (1890) 45:60
Nat Car (1890) 21:30
RR Gaz (1890) 22:14
Sci Am (1890) 62:26

DAB 1:193-4

Allen, John F. (1829-1900). Engineer and inventor of Allen engine, a pioneer in steam engines of the high speed type.

Am Mach (1900) 23:986,1210*
ASME Trans (1901) 22:1149
Eng News (1900) 44:396

DAB 1:199

Allen, John Howard (? -1898). Civil engineer in railroad work.

Eng Rec (1897-8) 37:491

Allen, Joseph (? -1900). Inventor of pneumatic riveter, high speed engines, and air compressor.

Eng Rec (1900) 42:354

Allen, Joseph Dana (1799?-1878). Civil engineer, canals and railroads.

RR Gaz (1878) 10:519

Allen, Richard N. (1825?-90). Locomotive engineer, inventor and manufacturer of paper car wheel.

Am Eng & RR J (1890) 64:522
Nat Car (1890) 21:170
RR Gaz (1890) 22:708

Allen, Theodore (1840-90). Mechanical and naval engineer; manufacturer of iron boats.

ASCE Proc (1891) 17:240-1

Allen, William Albert (1852-96). Civil engineer, Maine Central Railroad.

ASCE Proc (1896) 22:572
ASCE Trans (1896) 36:539
Assoc Eng Soc J (1897) 18:35-6, 55-6
Eng Rec (1895-6) 33:291
RR Gaz (1896) 28:224,561

Allen, William Daniel (1824?-96). Inventor of agitator for mixing steel.

Eng & Min J (1896) 62:467

Allis, Edward Phelps (1824-89). Manufacturer of steam
    and mill engines; builder of Reynolds Corliss engine.

    Am Eng & RR J (1889) 63:243
    Am Mach (4/11/1889) 12:8-9
    ASCE Proc (1889) 15:168-9*
    Eng Rec (1888-9) 19:250
    Ind World (4/4/1889) 32:5
    Iron Age (1889) 43:555,591*
    Power (5/1889) 9:13
    Sci Am (1889) 60:248

    DAB 1:219

Ames, Oliver (1831-95). Manufacturer and perfecter of
    shovels.

    Iron Age (1895) 56:913-4*

    DAB 1:254-6

Amsler, Carl (1848?-94). Engineer and contractor; con-
    structor of several iron and steel plants.

    AISA Bull (1894) 28:285
    Am Manuf (1894) 55:782
    Iron Age (1894) 54:1010*

Anderson, Adna (1827-89). Civil and military engineer;
    chief engineer of Northern Pacific Railroad.

    Am Eng & RR J (1889) 63:293
    ASCE Proc (1889) 15:166-8
    Eng & Min J (1889) 47:461
    Eng Rec (1888-9) 19:334
    Nat Car (1889) 20:94
    RR Gaz (1889) 21:331

Anderson, Charles Edward (1807-89). Civil engineer.

    Am Eng & RR J (1889) 63:531
    Eng & Min J (1889) 48:344

Anderson, Frank (? -1891). Inventor in the line of
    automatic telegraphy.

    Elec World (1891) 17:183

Anderson, James R. (? -1892). Military engineer; manu-
    facturer of cannons.

    Eng & Min J (1892) 54:276

Anderson, John F. (1823?-87). Civil engineer, railroads.

    RR Gaz (1887) 19:852

Anderson, Joseph R. (1813-92). Civil engineer and iron manufacturer.

Iron Age (1892) 50:841*

DAB 1:268-9

Andrews,Charles W. (? -1882). Civil engineer.

Am Eng & RR J (1882) 55:122

Andrews, Chauncey (1823-93). Manufacturer, mine operator, and railroad builder.

Eng & Min J (1893) 56:670
Ind World (12/28/1893) 41:7; (1/4/1894) 42:8
Iron Age (1894) 53:65*
RR Gaz (1894) 26:16

DAB 1:283-4

Andrews, Henry Porter (1822?-94). Civil engineer.

Eng & Min J (1894) 57:132
Eng Rec (1893-4) 29:168

Andrews, James (1824?-97). Constructing engineer and contractor.

Eng News (1897) 38:124*
RR Gaz (1897) 29:514

Andrews, James (1837?-97). Civil and mechanical engineer.

Eng & Min J (1897) 64:42
Iron Age (7/15/1897) 60:18

Angamar, Eugene H. (? -1880). Engineer and inventor in South.

Eng News (1880) 7:338

Anthony, Edward (1819-88). Civil engineer and photographer.

Eng Rec (1888-9) 19:54
Sci Am (1889) 60:49

Anthony, Henry T. (1814-84). Engineer, surveyor on Erie Road; improvements on Daguerre process of photography.

Sci Am (1884) 51:257

NYT (10/12/1884) 2:5

Appleton, Edward (? -1898). Civil engineer, planner of railroads in New England.

Eng Rec (1898) 38:201

Archbold, Samuel (1815?-90). Naval engineer, consulting engineer for railroads.

Am Eng & RR J (1890) 64:571
Iron Age (1890) 46:763

Arey, Clarence O. (1857?-96). Civil engineer.

Assoc Eng Soc J (1896) 17:14-5

Armstrong, Edward A. (1837?-91). Chief of public works department, Allegheny City, Pa.

Eng Rec (1890-1) 23:372

Arnold, Bishop (1853-87). Designer of traction machinery.

ASME Trans (1887) 8:727

Arnoldt, George (1820?-93). Civil engineer, division engineer of Erie Canal at Rochester.

Eng Rec (1892-3) 27:413

Ashburner, Charles Albert (1854-89). Mining engineer and geologist.

Am Eng & RR J (1890) 64:92
AIME Trans (1889) 18:365
ASME Trans (1890) 11:1153
Iron Age (1890) 45:13

Appleton's Cyc 1:109

Ashburner, William (1831-87). Mining engineer.

Am Eng & RR J (1887) 61:246
Eng & Min J (1887) 43:135

Ashton, Henry G. (1846-95). Inventor of Ashton safety valve for steam engines.

Am Mach (1895) 18:928
Eng & Min J (1895) 60:471
Eng Rec (1895) 32:435

Asmus, George (1830-92). Mining engineer.

Eng & Min J (1892) 53:659-60

Atkins, Jearum (1815-80). Inventor of Atkins' automaton
   (self rake), waterwheels, and other mechanical
   devices.

   Cassier's (1893-4) 5:109-12

   DAB 1:405-6

Attwood, Melville (? -1898). Assayer and metallurgist;
   inventor of method of amalgamating gold ores.

   Eng & Min J (1898) 65:528

Atwood, Luther (1826-69). Manufacturer of oil from coal.

   Sci Am (1869) 20:228

Atwood, William Henry (1857-90). Civil engineer, railroads.

   ASCE Proc (1891) 17:205-8

Avery, John (1830-84). Civil engineer, railroads and
   water works.

   ASCE Proc (1885) 11:17
   RR Gaz (1884) 16:97

Aycrigg, Benjamin (1804-95). Civil engineer, canals and
   coastal survey.

   Eng & Min J (1895) 59:131

Ayres, Bucher (? -1898). Civil engineer, railroads.

   RR Gaz (1898) 30:207

Babbitt, Benjamin Talbot (1809-89). Inventor and
   manufacturer of soap making machinery, boiler engine
   machinery, etc.

   Eng & Min J (1889) 48:364
   Iron Age (1889) 44:646

   DAB 1:455-6

Babcock, George H. (1832-93). Mechanical engineer, inventor of tubular boiler, steam engine, and polychromatic printing press.

Am Mach (12/28/1893) 16:8
Am Manuf (1894) 54:12
ASME Trans (1894) 15:636-9*
Elec Eng (1893) 16:536
Elec World (1894) 23:37
Eng (1890) 19:115*; (1893) 26:150
Eng & Min J (1893) 56:646
Eng Rec (1893-4) 29:52
Iron Age (1893) 52:1176*
Power (1/1894) 14:15
Sci Am (1893) 69:419

DAB 1:456-7

Bache, Alexander Dallas (1806-67). Superintendent of U.S. Coast Survey; researcher in terrestrial magnetism.

ASCE Proc (1896) 22:561-3
ASCE Trans (1896) 36:522-4
Eng & Min J (1866-7) 2:344
Frank Inst J (1869) 87:352-60
Pop Sci (1895-6) 48:112-20*

Appleton's Cyc 1:127*
DAB 1:461-3
NYT (2/20/1867) 2:2

Backus, Abner Lord (? -1895). Civil engineer, canals and railroads in Ohio.

Eng Rec (1895) 32:57

Bacon, Francis W. (1810?-86). Mechanical engineer.

Sci Am (1886) 54:98

Badger, Daniel D. (1806-84). Wrought iron manufacturer, pioneer in use of iron for building purposes.

Iron Age (11/27/1884) 34:17

NYT (11/19/1884) 2:3

Bailey, Charles Lukens (1821?-99). Iron manufacturer.

AISA Bull (1899) 33:157
Am Mach (1899) 22:876
Eng & Min J (1899) 68:314
Iron Age (9/7/1899) 64:17

Bailey, Jackson (1847-87). Editor, helped found American Machinist.

Am Eng & RR J (1887) 61:349
AISA Bull (1887) 21:203
Am Mach (7/30/1887) 10:8
Am Manuf (7/22/1887) 41:18
ASME Trans (1887) 8:728-9
Ind World (7/28/1887) 29:22
Nat Car (1887) 18:113
RR Gaz (1887) 19:476
Sanit Eng (1887) 16:158

Bailey, John A. (1819?-88). U.S. Government engineer, public works.

Am Eng & RR J (1888) 62:141
Sanit Eng (1887-8) 17:141

Bailey, R. T. (? -1873). Railroad engineer.

RR Gaz (1873) 5:54

Bailey, Thomas Norton (1850-86). Military engineer.

ASCE Proc (1890) 16:186-7

Baird, John (1820-91). Engineer and designer of steam-ships; worked on elevated railroads, iron works.

Eng (1891) 22:114
Iron Age (1891) 48:693
RR Gaz (1891) 23:754

Baird, Matthew (1817-77). Mechanic, manufacturer of locomotives.

AISA Bull (1877) 11:149
Iron Age (5/24/1877) 19:15
RR Gaz (1877) 9:233

DAB 1:511

Baker, L. C. (1839?-97). Superintendent of Western Union Company, constructed telegraph lines in South.

Elec World (1897) 29:464

Baker, Livingston L. (1827-92). Manufacturer of agricul-tural implements.

Eng & Min J (1892) 54:612

Baker, William L. (1850-88). Civil engineer, designer of iron bridges and roofs.

Am Eng & RR J (1888) 62:333
ASCE Proc (1889) 15:111-2
Assoc Eng Soc J (1888) 7:477-8
RR Gaz (1888) 20:375

Baker, W. S. Graff (? -1897). Electrical engineer.

Eng Rec (1897) 36:69

Baldwin, George Rumford (1798?-1888). Civil engineer, waterworks.

Am Eng & RR J (1888) 62:525
RR Gaz (1888) 20:694

Baldwin, Loammi (1780-1838). Surveyor and civil engineer, responsible for dry docks at Charleston and Newport.

Assoc Eng Soc J (1885-6) 5:10-25

Appleton's Cyc 1:149
DAB 1:540-1

Ball, C. E. (? -1897). Electrical inventor.

Elec Eng (1897) 23:67

Ball, Phinehas (1824-94). Civil engineer of Mass.

Assoc Eng Soc J (1896) 16:12
Eng Rec (1894-5) 31:57

Ball, Thomas (1824-94). Civil, consulting, and hydraulic engineer.

Iron Age (1894) 54:1172

Banfield, George (1815?-95). Manufacturer of first tin-plate in U.S.

AISA Bull (1895) 29:52

Barber, Dana C. (? -1889). Sanitary engineer, Phila-delphia.

Am Eng & RR J (1889) 63:388
Eng Rec (1889) 20:72

Barbour, William S. (1835-89). City engineer of Cambridge, Mass.; water works, sewers.

Am Eng & RR J (1889) 63:196
ASCE Proc (1889) 15:142
Assoc Eng Soc J (1889) 8:506-7
Eng Rec (1888-9) 19:180

Barker, W. P. (1829-99). Municipal engineer, railroads.

Eng Rec (1899) 40:37

Barkley, Rufus C. (? -1898). Locomotive engineer and
machinist.

Iron Age (8/11/1898) 62:18

Barnard, George A. (1841-99). Mechanical engineer,
Buckeye Engine Company.

Am Mach (1899) 22:139
ASME Trans (1899) 20:1008-9

Barnard, John Gross (1815-82). Military engineer,
supervised river and harbor improvements on major
rivers.

ASCE Proc (1887) 13:134-8
Eng News (1882) 9:162
Sci Am (1882) 46:381

Appleton's Cyc 1:169
DAB 1:626-7
NYT (5/15/1882) 5:5

Barnes, David Leonard (1858-96). Mechanical engineer,
designer of electric locomotives.

Am Eng & RR J (1897) 71:32
AISA Bull (1896) 30:285
Am Mach (1896) 19:1203
ASCE Proc (1899) 25:273-5
ASCE Trans (1899) 41:618-20
ASME Trans (1897) 18:1100-1
Elec Eng (1896) 22:653
Elec World (1896) 28:802
Eng & Min J (1896) 62:587
Eng Rec (1896-7) 35:47
Iron Age (1896) 58:1217
RR Gaz (1896) 28:898-9*
West Soc Eng J (1897) 2:100-1*

Barnes, James (1801?-69). Engineer and railroad contractor.

ASCE Proc (1896) 22:613-4
ASCE Trans (1896) 36:540-1

Appleton's Cyc 1:170-1
DAB 1:630-1

Barnum, William H. (1818-89). Manufacturer of pig iron and car wheels.

AISA Bull (1889) 23:124
Eng & Min J (1889) 47:418
Iron Age (1889) 43:705-6

Appleton's Cyc 1:173
NYT (5/1/1889) 5:5

Barr, Paul D. (? -1898). Civil engineer, railroads in West.

Eng Rec (1897-8) 37:403

Barrett, Amos R. (1833-95). Superintendent of motive power of Boston & Maine Railroad.

Nat Car (1895) 26:90
RR Gaz (1895) 27:320

Bartlett, Louis de B. (1825?-98). Mechanical engineer, inventor of mechanical improvements for steam engines and other machinery.

Am Mach (1898) 21:811
Iron Age (10/27/1898) 62:22

Bartlett, William Holmes Chambers (1809-93). Military and civil engineer, and mathematician; constructed coast fortifications; professor at West Point.

Eng & Min J (1893) 55:156
Eng Rec (1892-3) 27:231

Appleton's Cyc 1:186

Bartley, Joshua C. (1819-96). Mechanical engineer.

Iron Age (1896) 58:224

Bartol, Barnabas H. (1816-88). Mechanical engineer.

ASME Trans (1888) 9:737

Bassell, J. Y. (? -1899). Mining and metallurgical
engineer.

Eng Rec (1898-9) 39:266

Batchelder, Samuel (1784-1879). Inventor of dynamometer
used for ascertaining power for drying machinery.

Sci Am (1879) 40:128

Appleton's Cyc 1:191
DAB 2:40-1

Bates, Thomas C. (1812?-87). Builder of railroads in
Louisiana.

Sanit Eng (1887) 16:716

Bates, Thomas Henry (? -1893). Civil engineer, Erie
Canal, railroads.

RR Gaz (1893) 25:177

Battin, Joseph (1806?-93). Mechanical engineer, inventor
of coal breaker.

Eng & Min J (1893) 56:246

Bauer, Alexander H. (1846-95). Chief electrician of
Pullman Palace Company; developed practical secondary
battery applications for light and power.

AIEE Trans (1894) 11:874-6; (1895) 12:667
RR Gaz (1895) 27:60

Bauer, Charles A. (1847-99). Mechanical engineer.

Am Mach (1899) 22:568*,571
ASME Trans (1899) 20:1017

Bayley, George William Redd (1821-76). Civil engineer,
hydraulic works and railroads.

ASCE Proc (1878) 4:58-60

Beach, Alfred Ely (1826-96). Inventor of pneumatic
devices; editor of Scientific American.

Am Eng & RR J (1896) 70:26
Elec Eng (1896) 21:49
Eng & Min J (1896) 61:36*
Eng Rec (1895-6) 33:75
Iron Age (1896) 57:147
RR Gaz (1896) 28:30
Sci Am (1896) 74:18-9*

DAB 2:80-1

14

Beach, Charles A. (? -1899). Civil engineer of western N.Y. State.

Eng Rec (1899) 40:658

Beach, Moses Sperry (1822-92). Inventor and journalist.

Sci Am (1892) 67:80
DAB 2:81-2

Beach, Nelson (? -1876). Railroad and canal engineer.

RR Gaz (1876) 8:99

Beale, Horace A. (1827-97). Iron manufacturer.

AISA Bull (1897) 31:253
Eng & Min J (1897) 64:552
Eng Rec (1897) 36:487

Beardsley, Frank (1860-95). Civil and mining engineer, railroads.

ASCE Proc (1895) 21:182-3

Beatty, James (? -1893). Mechanical engineer.

Eng Rec (1893) 28:262

Becker, Albrecht (1821-92). Mechanical engineer, iron works and water works.

Eng (1893) 25:2
Eng Rec (1892-3) 27:92
Iron Age (1892) 50:1281

Becker, Max Joseph (1827-96). Railroad and bridge engineer.

ASCE Proc (1897) 23:124
ASCE Trans (1897) 37:555-7
Eng & Min J (1896) 62:203
Eng News (1889) 21:126-7
Eng Rec (1896) 34:234
Iron Age (1896) 58:413
RR Gaz (1896) 28:610

Bedell, Charles Edward (1863-1900). Bridge architect and engineer; built several structures in N.Y. City.

AISA Bull (1900) 34:173
Eng & Min J (1900) 70:394
Iron Age (10/4/1900) 66:25
RR Gaz (1900) 32:661

Beggs, James (1843?-89). Mechanical and consulting engineer; dealer in stationary engines and machinery.

Am Eng & RR J (1889) 63:388
Am Mach (7/25/1889) 12:7
ASME Trans (1889) 10:832
Eng Rec (1889) 20:114
Iron Age (1889) 44:132

Behrens, William Frederick (1860-94). Civil engineer, railroads in Mexico and West.

ASCE Proc (1894) 20:87-8

Belden, N. M. (? -1895). Civil engineer.

Eng Rec (1895) 32:183

Belknap, M. K. (? -1890). Civil engineer, railroads.

Am Eng & RR J (1890) 64:428

Belknap, Morris Sheppard (1845-90). Draftsman; developer of stone quarry at Bowling Green, Ky.

ASCE Proc (1890) 16:167-8

Bell, James E. (1849-79). Civil engineer, railroads, locks, and dams.

ASCE Proc (1879) 5:98-100

Bellhouse, R. W. (1856-98). Mechanical engineer.

ASME Trans (1899) 20:1014
Eng Rec (1897-8) 37:403

Bellman, Oscar (? -1897). Inventor of steam and hot water heating apparatus.

Eng Rec (1897) 36:69

Bement, William Barnes (1817-97). Manufacturer of machine tools.

Am Eng & RR J (1897) 71:387
AISA Bull (1897) 31:229
Am Mach (1897) 20:783
Am Manuf (1897) 61:555
RR Gaz (1897) 29:735

Benbow, P. J. (? -1897). Mining engineer, president of Otis Steel Company, Cleveland.

Eng Rec (1897) 36:509

Bender, Andrew Sigourney (1820-97). Civil engineer and surveyor of Calif.

Eng & Min J (1897) 64:732
Eng Rec (1897-8) 37:47

Bennett, Edwin Howard (1845-98). Mechanical engineer and inventor of details for Singer sewing machine.

Am Mach (1898) 21:511
ASME Trans (1898) 19:979-80

Bennett, Ensign (1831-88). Resident engineer of Erie Canal; builder of Valley Canal Railroad.

Am Eng & RR J (1888) 62:190
Eng & Min J (1888) 45:148
Sanit Eng (1887-8) 17:204

Bennett, Jeremiah (? -1881). Builder of first wooden railroad west of the Appalachians.

RR Gaz (1881) 13:57

Bentley, H. A. (? -1889). City engineer of Newport, R.I.

Eng Rec (1888-9) 19:348

Bentley, Henry (1834-95). "Father" of the local telegraph system.

Elec Eng (1895) 20:289

Benton, John Dean (1823-90). Inventor and model maker.

Eng & Min J (1890) 50:485
Iron Age (1890) 46:721

Benyaurd, William H. H. (? -1900). Builder of bridges in Mississippi Valley.

Eng Rec (1900) 41:139

Berdan, Hiram (1823-93). Inventor of Berdan range finder, torpedo, and rifle.

Eng & Min J (1893) 55:324
Iron Age (1893) 51:792

NYT (4/1/1893) 1:6

Beresford, Frank (1861-87). Civil and railroad engineer.

ASCE Proc (1896) 22:700
ASCE Trans (1896) 36:594.

Berger, John Jacob (1865-99). Electrical engineer.

Elec World (1899) 34:764

Bergner, Theodore (1844-86). Mechanical engineer and
inventor.

ASME Trans (1886) 7:829-30
RR Gaz (1886) 18:29

Bigelow, Erastus Brigham (1814-79). Inventor of automatic
carpet loom.

Sci Am (1879) 41:409

Appleton's Cyc 1:260
DAB 2:254-5

Billings, Albert M. (1814-97). Pioneer in use of
illuminating gas and elevated railroads.

Eng Rec (1896-7) 35:223

Bingham, John F. (? -1891). Naval engineer.

Eng Rec (1890-1) 23:372

Birkinbine, Henry P. M. (1819-86). Hydraulic engineer.

Am Manuf (5/7/1886) 38:9
Frank Inst J (1886) 122:302-5
Iron Age (4/29/1886) 37:18

Bishop, Thomas Sparks (1846-98). Civil engineer, railroads in N.J.

ASCE Proc (1899) 25:330
ASCE Trans (1899) 41:621

Bispham, Servetus (? -1897). Railway and locomotive
engineer; constructor of foundations for several
buildings for Chicago World's Fair.

Eng Rec (1897-8) 37:139

Bissell, Levi (? -1873). Inventor of railroad machinery
and Bissell truck.

RR Gaz (1873) 5:333

Bixby, Edgar M. (1847-92). Draftsman of Boston.

ASME Trans (1892) 13:679-80

Black, Alexander L. (1824?-99). Mechanical engineer.

Am Mach (1899) 22:289

Blackstone, Timothy B. (1829-1900). Civil engineer, railroads.

RR Gaz (1900) 32:363

Blake, Edward (1862-93). Electrical engineer.

Elec Eng (1893) 16:377

Blake, Eli Whitney (1795-1886). Inventor of Blake stone crusher, locks, casters, hinges, and firearms.

Eng & Min J (1886) 42:135
Eng Rec (1895) 32:345
Iron Age (8/26/1886) 38:23
Pop Sci (1886-7) 30:432
Sci Am (1886) 55:127

Appleton's Cyc 1:283
DAB 2:341-2

Blake, Francis C. (1854-91). Machinist; superintendent of Pennsylvania Lead Company.

ASME Trans (1891) 12:1055-6

Blake, George W. (1842?-90). Manufacturer of steam heating and ventilating apparatus.

Eng Rec (1889-90) 21:317
Iron Age (1890) 45:647

Blake, J. H. (1800?-99). Chemist and mining engineer.

Iron Age (7/13/1899) 64:19

Blanc, Amidee (1809?-39). Civil engineer.

Am Eng & RR J (1839) 9:132-3

Blanchard, Albert (1810-91). Civil engineer and surveyor.

Eng & Min J (1891) 51:748
Eng Rec (1891) 24:52

Appleton's Cyc 1:287

Blanchard, Thomas (1788-1864). Inventor of machinery for the manufacture of tacks and for turning and finishing gun barrels.

Sci Am (1864) 10:282; (1869) 21:395

Appleton's Cyc 1:288
DAB 2:351-2
NYT (4/24/1864) 3:5

Bleloch, George H. (1835-91). Mechanical engineer, manufacturer of needles for sewing machines.

ASME Trans (1892) 13:673-5

Bliss, Henry Isaac (1830-96). Railroad engineer, city engineer of La Crosse, Wisc.

ASCE Proc (1896) 22:135
ASCE Trans (1896) 36:541

Blood, Aretas (1816?-97). Machinist, established and conducted locomotive works.

Am Eng & RR J (1898) 72:23
Am Mach (1897) 20:913
RR Gaz (1897) 29:859

Blunden, Henry D. (1849-89). Railroad engineer.

ASCE Proc (1896) 22:575
ASCE Trans (1896) 36:542

Blunt, Charles E. (1823?-92). Military engineer, fortifications of rivers and harbors.

Am Eng & RR J (1892) 66:385
Eng Rec (1892) 26:106

Blythe, Washington (1809-82). Civil engineer.

RR Gaz (1882) 14:171

Boardman, Napoleon (1825-99). Engineer, railways.

Eng Rec (1899) 40:734

Bodfish, Sumner Homer (1844-94). Hydraulic engineer; topographer and irrigation engineer for U.S. Geological Survey.

ASCE Proc (1894) 20:96-8

Boeke, Augustus W. (1860-94). Municipal engineer, Kansas.

ASCE Proc (1894) 20:204

Boericke, Rudolph (1864?-97). Mechanical and mining engineer.

Eng & Min J (1898) 65:78
Iron Age (1/20/1898) 61:20

Bogardus, James (1800-74). Inventor of machinery; builder of first cast iron building in the world in N.Y. City.

Iron Age (4/16/1874) 13:15
Sci Am (1869) 21:394; (1874) 30:276

Appleton's Cyc 1:301
DAB 2:407-8

Bogart, Abraham L. (1818-96). Inventor and gas engineer.

Ind World (8/13/1896) 47:5

Bole, Hugh M. (1826-1900). Manufacturer of machines and guns.

AISA Bull (1900) 34:213
Iron Age (12/13/1900) 66:43

Bollman, Wendel (1814?-84). Engineer and bridge builder.

RR Gaz (1884) 16:211

Bond, M. B. (? -1898). Engineer of South, superintendent of construction at Fort Morgan; railroads in Alabama.

Eng Rec (1897-8) 37:491

Bonnell, Henry O. (1839-93). Iron and steel manufacturer of Youngstown, Ohio.

AISA Bull (1893) 27:43
Am Manuf (1893) 52:103
Eng Rec (1892-3) 27:151
Iron Age (1893) 51:136,253*

Booker, Bernard Frank (1858-94). Railroad engineer of Southwest, Midwest, and Mexico.

ASCE Proc (1894) 20:183-5

Boomer, L. B. (1820?-81). Civil engineer, president of American Bridge Company.

Am Eng & RR J (1881) 54:233
RR Gaz (1881) 13:147
Sci Am (1881) 44:193

NYT (3/7/1881) 5:2

Booth, Edgar Hubbard (1861-97). Mechanical engineer.

ASME Trans (1898) 19:974-5

Booth, Kirtland Farnum (1829-92). Railroad engineer.

Assoc Eng Soc J (1893) 12:220-1

Borda, Eugene (1825-97). Engineer, owner of coal mines.

Eng & Min J (1897) 63:401*

Botlicher, Otto (? -1897). Civil engineer, railroads.

Eng Rec (1896-7) 35:179

Bowden, J. H. (1846-1900). Chief engineer of anthracite coal mining company; inventor of self-oiling car wheel.

Eng & Min J (1900) 70:615

Bowen, Edmund S. (1831-97). Railroad engineer.

Am Eng & RR J (1897) 71:317
RR Gaz (1897) 29:610

Bowen, James L. (? -1898). Engineer for Confederacy during Civil War.

Eng Rec (1897-8) 37:139

Bowen, Menard K. (1858-99). Railway engineer.

Elec World (1899) 33:492
Eng Rec (1898-9) 39:457
RR Gaz (1899) 31:270

Bowers, Edwin L. (? -1897). Civil engineer, railroads.

Eng Rec (1897) 36:245

Bowron, James (? -1877). Mining and civil engineer; organized Southern States Coal, Iron and Land Company.

Iron Age (12/6/1877) 20:5

Boyd, James A. (? -1895). Inventor of brickmaking machinery, manufacturer of bricks.

Eng Rec (1895-6) 33:93

Boyd, N. W. (1844?-1900). Inventor of switch apparatus.

Am Mach (1900) 23:146

Boyden, Seth (1788-1870). Inventor and manufacturer; produced first "patent" leather, later produced iron locomotives.

Sci Am (1870) 22:238

Appleton's Cyc 1:341
DAB 2:529

Boyden, Uriah Atherton (1804-79). Hydraulic engineer;
   "father" of American mixed-flow hydraulic turbine
   design.
   Sci Am (1879) 41:296
   Appleton's Cyc 1:341
   DAB 2:529-30

Boyer, Zaccur Prall (1832-1900). Inventor; manufacturer
   of iron and street cars.
   AISA Bull (1900) 34:197
   Eng Rec (1900) 42:500
   Iron Age (11/12/1900) 66:44

Boyle, James E. (1834?-90). Inventor of sanitary appliances.
   Eng Rec (1890-1) 23:2

Boyne, M. W. (? -1897). Civil engineer, railroads.
   Eng Rec (1896-7) 35:487

Bracher, Thomas W. (1843?-99). Inventor of machines in
   hat industry.
   Am Mach (1899) 22:1141

Bradley, Alfred (? -1897). Civil engineer and architect.
   Eng Rec (1897) 36:509

Bradley, David (1812-99). Manufacturer of agricultural
   implements, started the first foundry in Chicago.
   Am Mach (1899) 22:177
   Iron Age (3/2/1899) 63:22

Bradley, George (? -1898). Pioneer railroad builder
   in West.
   Eng Rec (1897-8) 37:183

Bradley, Leverett (1798?-1875). Electrical inventor.
   Sci Am (1875) 33:223

Bradley, Osgood (? -1896). Car builder.
   RR Gaz (1897) 29:34

Bradley, Sylvanus (1833-91). Master mechanic.
   Nat Car (1891) 22:140

Bramwell, J. Herbert (1846-94). Mining engineer.

AIME Trans (1894) 24:749-51
AISA Bull (1894) 28:162
Am Manuf (1894) 55:161
Eng & Min J (1894) 58:59
Eng Rec (1894) 30:118
Iron Age (1894) 54:107
RR Gaz (1894) 26:513

Brandt, John (? -1890). Civil engineer, railroads.

Am Eng & RR J (1890) 64:428
Eng & Min J (1890) 50:132

Bray, R. T. (? -1897). Mechanical engineer.

Eng Rec (1897) 36:333

Brazier, James (? -1889). Steamboat engineer.

Eng (1889) 17:14

Breckenridge, J. M. (? -1896). Inventor of improvements in clocks and clockmaking machinery.

Am Mach (1896) 19:1167

Breed, Gilbert C. (1829-86). Railroad engineer.

Am Eng & RR J (1887) 61:39
RR Gaz (1886) 18:827

Breitung, Edward (1831-87). Responsible for development of Lake Superior iron.

Am Eng & RR J (1887) 61:150
Iron Age (3/17/1887) 39:21

Brenneke, Charles (? -1893). Engineer and architect.

Eng Rec (1893) 28:102

Brevoort, Henry Lefferts (1849?-95). Mechanical engineer and patent expert.

Am Eng & RR J (1895) 69:379
Eng Rec (1895) 32:111
Iron Age (1895) 56:72

Brevoort, James Carson (1818-87). Civil engineer, surveyor of Northeast boundary.

Am Eng & RR J (1888) 62:46
Eng & Min J (1887) 44:436
Sanit Eng (1887-8) 17:28

Appleton's Cyc 1:369

Brewer, F. B. (1820-92). Pioneer in petroleum industry.

   Eng & Min J (1892) 54:132

Brewer, Paul C. (? -1899). Engineer of Pennsylvania
   Railroad.

   RR Gaz (1899) 31:107

Brewerton, Henry (1801-79). Military engineer.

   Eng News (1879) 6:122

   Appleton's Cyc 2:370
   NYT (4/18/1879) 5:2

Brewster, Benjamin (1828-97). Constructor of elevated
   railroads in N.Y.

   RR Gaz (1897) 29:640

Briggs, Albert Dwight (1820-81). Civil engineer and
   railroad commissioner of Mass.

   ASCE Proc (1889) 15:132-3
   RR Gaz (1881) 13:117

Briggs, Robert (1822-82). Mechanical engineer and expert
   on ventilation; designed plans for Capitol building.

   ASCE Proc (1896) 22:567-9
   ASCE Trans (1896) 36:542-5
   Frank Inst J (1882) 114:240; (1883) 115:229
   Iron Age (8/3/1882) 30:28
   RR Gaz (1882) 14:462
   Sanit Eng (1882) 6:182,342
   Sci Am (1882) 47:116

Brill, John George (1817-88). Manufacturer of railway
   cars.

   Iron Age (1888) 42:468
   RR Gaz (1888) 20:643

Bringhurst, John H. (1822?-98). Saw manufacturer of
   Philadelphia.

   AISA Bull (1899) 33:4
   Am Mach (1899) 22:18

Brink, Mahlon S. (1822?-95). Established first anthracite blast furnace in U.S.

   Eng & Min J (1895) 60:543
   Iron Age (1895) 56:1104

Brinsmade, William B. (? -1880). Civil engineer, railroads.

Eng News (1880) 7:184

Bristol, M. C. (1842-97). Superintendent of construction, Western Union.

Elec Eng (1897) 23:482
Elec World (1897) 29:602

Britton, Hiram M. (1831-89). Mechanic and engineer; general manager of railroads.

RR Gaz (1889) 21:546

Bromley, William (? -1890). Builder of marine engines.

Iron Age (1890) 46:955

Brooks, David (1820?-91). Inventor and electrician, helped install first telegraph line in America.

Am Eng & RR J (1891) 65:331
Elec Eng (1891) 11:639
Eng Rec (1891) 24:2
Frank Inst J (1891) 132:145-7

Brooks, Horatio G. (1828-87). Machinist, locomotive works.

Am Eng & RR J (1887) 61:200
RR Gaz (1887) 19:293

Brooks, John W. (1818?-81). Civil engineer, railroads.

RR Gaz (1881) 13:528

Brooks, Thomas Benton (1836-1900). Civil and mining engineer, geologist.

Eng & Min J (1900) 70:645
Eng Rec (1900) 42:524

Appleton's Cyc 1:390
DAB 3:89-90

Brown, Charles P. (? -1900). Professor of mining and metallurgy in Nevada.

Eng & Min J (1900) 70:135

Brown, Edward Dexter (1868-98). Electrical engineer, general inspector of AT&T.

Elec Eng (1898) 26:85
Elec World (1898) 32:127
Eng Rec (1898) 38:179

Brown, Ezra (1827?-80). Master car builder.

Nat Car (1880) 11:179*,181

Brown, F. C. (? -1898). Engineer of Brattle Block, Boston.

Am Mach (1898) 21:848

Brown, Felix (1826-99). Inventor of pulleys and shafts.

Am Mach (1899) 22:311
Iron Age (4/13/1899) 63:18

Brown, Harvey (1793?-1871). Inventor of lamps and method of propelling street cars by wire rope.

Sci Am (1871) 24:9

Brown, Hiram W. (? -1862). Inventor of improved cotton gin.

Sci Am (1862) 7:51

Brown, John Milton (1845-74). Civil engineer, railroads.

ASCE Proc (1873-5) 1:170-1

Brown, Myron E. (1827?-79). Locomotive engineer.

RR Gaz (1879) 11:651

Brown, R. H. (? -1899). Chief engineer of Delaware & Hudson Canal Company.

Eng Rec (1898-9) 39:409

Brown, Robert Newland (1816?-86). Civil engineer, railroads.

RR Gaz (1886) 18:857

Browne, Spencer C. (? -1896). Mining engineer.

Eng & Min J (1896) 62:539

Bruce, David (1802?-92). Inventor of typecasting machine.

Eng & Min J (1892) 54:301

Brunner, Burroughs Price (1829?-81). Engineer and inventor of machinery for utilizing old steel rails.

Sci Am (1881) 45:38

Brunot, Felix R. (1820-98). Civil engineer; iron and steel manufacturer.

AISA Bull (1898) 32:77

Appleton's Cyc 1:419

Brush, Charles Benjamin (1848-97). Civil engineer, waterworks projects.

ASME Trans (1898) 19:967
Eng Rec (1897) 36:3
RR Gaz (1897) 29:424

Brush, William P. (1839?-90). Mechanical engineer.

Eng & Min J (1890) 50:551
Iron Age (1890) 46:854

Bruyn, Levi D. (1835-87). Civil engineer, railroads.

Am Eng & RR J (1887) 61:200
RR Gaz (1887) 19:258

Bucey, John H. (? -1896). Civil engineer, employed in mines.

Eng Rec (1896-7) 35:47

Bucke, Maurice A. (1868-99). Mining engineer.

AIME Trans (1900) 30:xxv

Buckhout, Isaac Craig (1831-74). Civil engineer, chief engineer of New York & Harlem Railroad; designer of Grand Central Station, N.Y. City.

ASCE Proc (1873-5) 1:171-2
RR Gaz (1874) 6:387
Sci Am (1874) 31:258

Appleton's Cyc 1:437
DAB 3:226-7

Buckland, Cyrus (1799-1891). Inventor of parts for fire-arms, and designer of machinery and tools for their manufacture.

Iron Age (1891) 47:440

Appleton's Cyc 1:439
DAB 3:229

Bucklin, James M. (1801?-90). Civil engineer, state engineer of Illinois.

Eng Rec (1889-90) 21:338

Buestrin, Henry (? -1893). Mechanical engineer, known for raising and moving buildings.

Eng Rec (1892-3) 27:271

Bullock, Milan C. (1838-99). Manufacturer of mining machinery; developer of diamond drill for use in prospecting.

AIME Trans (1900) 30:xxv-xxvi
Am Mach (1899) 22:59-60
ASME Trans (1899) 20:1006-7
Elec Eng (1899) 27:99
Elec World (1899) 33:98*
Eng & Min J (1899) 67:43-4*
Iron Age (1/14/1899) 63:17
RR Gaz (1899) 31:51
West Soc Eng J (1899) 4:239-40*

Burden, Henry (1791-1871). Inventor and manufacturer of horseshoe machine, cultivator, and other agricultural implements.

Sci Am (1869) 21:409; (1871) 24:73

DAB 3:272
NYT (1/21/1871) 8:5

Burgess, Edward (1848-91). Ship designer.

Am Eng & RR J (1891) 65:381
Sci Am (1891) 65:55,129

Appleton's Cyc 1:451
NYT (7/13/1891) 2:4

Burleigh, Charles (1824?-83). Inventor of rock drill.

RR Gaz (1883) 15:371

Burlingame, Abraham (1842-1900). Inventor and mechanical engineer; builder of steam engines.

Iron Age (2/22/1900) 65:20

Burlingame, J. S. (1838?-99). Military and railroad engineer.

Eng Rec (1899) 40:658

Burnet, Samuel Forder (1860-91). Mining engineer.

Assoc Eng Soc J (1891) 10:262-4
Eng Rec (1890-1) 23:188

Burnett, J. H. (1826-85). Mechanical engineer.

ASME Trans (1885) 6:872

Burnham, Arthur H. (? -1877). Captain of engineers, U.S.

Eng News (1877) 4:261

Burr, James Dewey (1843-86). Civil engineer, railroads and bridges.

ASCE Proc (1886) 12:107-8

Burr, Shields (? -1883). Civil engineer.

RR Gaz (1883) 15:714

Burroughs, William (1851-98). Inventor of adding machine.

Am Mach (1898) 21:752

Burrowes, Richard W. (1823?-97). Civil engineer, railroads.

Eng & Min J (1897) 63:516

Burrows, George H. (1822?-96). Railroad engineer, superintendent of Western division of N.Y. Central.

RR Gaz (1896) 28:187

Burt, Austin (1818?-94). Inventor of solar compass; president of Detroit Transit Railway Company.

Iron Age (1894) 53:420

Burt, William (1825?-98). Inventor of solar compass
and typograph (typewriting machine); surveyor.

Eng & Min J (1898) 66:764

Burtis, Divine (1811-87). Steamboat builder.

Ind World (9/22/1887) 29:1

Burton, Thomas (? -1900). Mechanical engineer.

Am Manuf (1900) 66:47

Bushnell, Cornelius Scranton (1828-96). Inventor and
railroad builder; one of initiators of first transcontinental railroad.

Eng & Min J (1896) 61:451
Iron Age (1896) 57:1139
NYT (5/7/1896) 1:3

Butterfield, Frederick E. (1863-85). Machinist
and draftsman.

ASME Trans (1886) 7:827

Butterworth, James (? -1891). Naval engineer, authority
on modern machinery.

Am Eng & RR J (1891) 65:524
Eng (1891) 22:88

Butterworth, Richard Edward Emerson (1806-88). Manufacturer of pumping engines and machinery for Grand
Rapids water works.

Sanit Eng (1887-8) 17:124

Button, Lysander (1810?-98). Builder of steam fire
engines.

Am Mach (1898) 21:625

Butts, David M. (? -1897). Civil engineer.

Eng Rec (1896-7) 35:377

Butts, Elijah Polhill (1856-92). Civil engineer, bridges.

ASCE Proc (1892) 18:129-31

Byam, S. T. J. (? -1895). Mechanic engaged in manufacture of watches and fine interchangeable mechanisms.

Am Mach (1895) 18:48

Byers, Alexander McBurney (1827-1900). Manufacturer of wrought iron pipe.

AISA Bull (1900) 34:165
Am Manuf (1900) 67:239*
Iron Age (9/27/1900) 66:29
RR Gaz (1900) 32:643

Byers, John Morton (1832?-93). Engineer, railroads in Pa.

Eng & Min J (1893) 55:252
RR Gaz (1893) 25:216

Byers, Joseph (? -1883). Civil engineer, railroads.

RR Gaz (1883) 15:238

Cahill, John (? -1896). Steam engineer, president of company of machinists and boiler makers.

Eng Rec (1896-7) 35:91

Caldwell, D. W. (1830-97). Civil engineer.

Am Eng & RR J (1897) 71:283

Callahan, Denis (? -1896). Draftsman and topographical engineer.

Eng Rec (1896) 34:439

Cameron, Adam Scott (? -1877). Manufacturing engineer, constructor of steam pumps.

Sci Am (1877) 37:353

Camp, Jonathan (1838-74). Civil engineer, railroads in N.Y.; public works in N.J.

ASCE Proc (1873-5) 1:186

Campbell, Allan (1815-94). Civil engineer, constructor of railroads in South America and N.Y. State.

ASCE Proc (1894) 20:179-82
Eng & Min J (1894) 57:276
Eng News (1894) 31:237
Eng Rec (1893-4) 29:264
RR Gaz (1894) 26:220

DAB 3:448-9

Campbell, Andrew (1821-90). Inventor and manufacturer of printing presses.

Eng & Min J (1890) 49:453

DAB 3:449-50

Campbell, Galvin (1836-94). Master mechanic, Wisc.

Nat Car (1894) 25:44

Campbell, John (1808-91). Manufacturer of iron beam plow.

AISA Bull (1891) 25:276
Am Manuf (1891) 49:511
Iron Age (1891) 48:500

Campbell, John C. (1817-90). Hydraulic engineer; worked on Croton Reservoir, Hudson River Railroad, and several public works in N.Y. City.

Am Eng & RR J (1890) 64:237
Eng & Min J (1890) 49:365
Eng Rec (1889-90) 21:258
Iron Age (1890) 45:557-8
RR Gaz (1890) 22:221

Campbell, Robert B. (1832-1900). One of the builders of the Hoosac Tunnel.

Eng Rec (1900) 42:258

Canby, Samuel (? -1897). Civil and park engineer of Wilmington, Del.

Eng Rec (1897) 36:201

Canty, Stephen Montague (? -1897). Electrician and inventor.

Elec World (1897) 30:206

Card, Joseph (1837-94). Inventor in the line of wood preservation techniques.

ASCE Proc (1895) 21:68

Carnell, George W. (1832?-99). Manufacturer of brick making machinery.

Am Mach (1899) 22:650

Carney, Frank J. (? -1897). Mechanical engineer.

Eng Rec (1897) 36:91

Carpenter, Clarence A. (1846-99). Railroad engineer of Mich.

Eng Rec (1899) 40:563
RR Gaz (1899) 31:799

Carpenter, James H. (1847?-98). Inventor of Carpenter projectile; steel manufacturer.

AISA Bull (1898) 32:45
Am Mach (1898) 21:192
Am Manuf (1898) 62:337
Eng & Min J (1898) 65:318
Eng Rec (1897-8) 37:315
Ind World (3/10/1898) 50:6
Iron Age (3/10/1898) 61:18

Carrell, Frederick Janvrin (1852-94). Civil engineer, canals, sewers, and other water works in West.

ASCE Proc (1897) 23:148
ASCE Trans (1897) 37:559

Carrington, John Warner (? -1895). Civil engineer.

Eng Rec (1894-5) 31:381

Carson, John B. (1832?-92). Railroad engineer and manager.

Nat Car (1892) 23:31

Carter, Frank (1834-99). Mining engineer.

AIME Trans (1900) 30:xxviii
AISA Bull (1899) 33:61
Iron Age (3/30/1899) 63:18

Carter, Marshal W. (1825-90). Civil engineer.

Iron Age (1890) 45:605

Cartwright, Henry (1823-81). Civil engineer and machinist, worked on Hoosac Tunnel.

Am Eng & RR J (1881) 54:773
ASCE Proc (1881) 7:124-5
Frank Inst J (1881) 112:399

Cartwright, James (1828?-92). Iron manufacturer.

AISA Bull (1892) 26:285
Am Manuf (1892) 51:585
Eng & Min J (1892) 54:348

Case, Jerome Increase (1818-91). Manufacturer, designer, and builder of power threshers.

Iron Age (1891) 48:1170

DAB 3:556

Case, William H. (? -1898). Civil and mining engineer.

AIME Trans (1899) 29:xxvi
Eng & Min J (1898) 66:554
Eng Rec (1898) 38:465

Casey, Thomas Lincoln (1831-96). U.S. Army Chief of Engineers; responsible for several major buildings in Washington, D.C., including the Library of Congress.

Eng & Min J (1896) 61:307
Eng Rec (1895-6) 33:291,309
RR Gaz (1896) 28:241
Sci Am (1896) 74:211

Casey, William R. (1808?-46). Civil engineer, railroads.

Am Eng & RR J (1846) 19:537

Casgrain, William T. (1835-1900). Civil engineer, surveyor, and contractor; built several waterworks structures.

West Soc Eng J (1900) 5:571-2

Cass, George Washington (1810-88). Civil engineer, dams and railroads.

ASCE Proc (1896) 22:704-6
ASCE Trans (1896) 36:599-602
RR Gaz (1888) 20:195
Sanit Eng (1887-8) 17:258

DAB 3:561-2

Cassidy, Patrick (? -1896). Manufacturer of plumbers' materials and steam boilers.

Eng Rec (1896) 34:271

Cassin, Isaac S. (? -1897). Civil and hydraulic engineer; public building commissioner of Philadelphia.

Eng Rec (1896-7) 35:311
Iron Age (2/11/1897) 59:17

Castner, Hamilton Young (1859?-99). Electrolytical chemist and inventor.

Eng & Min J (1899) 68:485-6*
Iron Age (10/19/1899) 64:17

Caton, William B. (? -1896). Railroad engineer in U.S. and South America.

Eng Rec (1896) 34:307

Cauldwell, James A. (1840?-98). Inventor of agricultural implements and lawnmower; iron manufacturer.

Iron Age (3/17/1898) 61:27

Cavner, A. R. (? -1891). Locomotive engineer.

Nat Car (1891) 22:172

Cawley, Frank (1868-96). Mechanical engineer.

ASME Trans (1896) 17:746

Chabot, Cyprien (1824-89?). Toolmaker and inventor of firearms, sewing machine, shoe manufacturing machine, etc.

Frank Inst J (1890) 129:68

Chalfant, John W. (1827-98). Iron and steel manufacturer, Pittsburgh.

AIME Trans (1899) 29:xxvii-xxviii
AISA Bull (1899) 33;4
Am Mach (1899) 22:18
Am Manuf (1898) 63:951
Eng & Min J (1898) 66:794
Eng Rec (1898-9) 39:128
Iron Age (1/5/1899) 63:21*

Chalfin, Samuel F. (1826?-91). Civil engineer.

Eng Rec (1891) 24:360

Chapin, Charles L. (1830?-96). Electrician, one of the early telegraph operators.

Elec Eng (1896) 22:307
Elec World (1896) 28:378
Eng Rec (1896) 34:289

Chapman, George (1870-1900). Mechanical engineer.

Iron Age (6/7/1900) 65:39

Chapman, Herbert W. (1848-99). Manufacturer of engraving engines for bank note and bond work.

Am Mach (1899) 22:97
Iron Age (2/2/1899) 63:26

Chapman, J. C. (1822-98). Machinist and inventor of
Chapman valve.

Am Mach (1898) 21:867
Eng Rec (1898) 38:509
Iron Age (11/10/1898) 62:17

Chapman, Luke (1835-91). Improver of machines for
machine forging and metal working.

ASME Trans (1891) 12:1057-8

Chase, Samuel Stewart (1825-73). Hydraulic engineer.

ASCE Proc (1873-5) 1:40-1

Chase, William Livingston (1855-98). Mechanical engineer.

ASME Trans (1899) 20:1002

Cheever, Charles A. (1852-1900). Inventor of improvements on telephone.

Elec World (1900) 35:722

Cheney, Orlando H. (1839-94). Civil engineer, superintendent of sewers in Chicago.

Assoc Eng Soc J (1895) 14:19-20
Eng Rec (1893-4) 29:328

Cheney, Ward (1813-76). Pioneer manufacturer of silk.

AISA Bull (1876) 10:124

DAB 4:56
NYT (3/23/1876) 2:4

Cherry, E. V. (? -1899). Manufacturer of electrical and
telephone apparatus.

Elec World (1899) 34:1029

Cherry, James M. (? -1897). Mining engineer and railroad contractor.

Eng Rec (1896-7) 35:135

Chesbrough, Ellis Sylvester (1813-86). City engineer of
Boston and Chicago; authority on water supply and
sewerage of cities.

ASCE Proc (1889) 15:160-3
Assoc Eng Soc J (1886-7) 6:129
Eng News (1886) 16:123-4
Iron Age (8/26/1886) 38:23
Nat Car (8/26/1886) 17:5
RR Gaz (1886) 18:601

Appleton's Cyc 1:599

Chesbrough, Isaac Collins (1814-93). Civil and railroad
engineer.

Assoc Eng Soc J (1894) 13:145-7
Eng & Min J (1893) 55:204
Eng Rec (1892-3) 27:211
RR Gaz (1893) 25:100

Chester, Charles T. (1840-80). Inventor of fire alarm
telegraph.

Eng News (1880) 7:136
NYT (4/14/1880) 4:7

Chester, Stephen M. (? -1894). Civil engineer and
electrician; inventor of electrical sewing machine
motor and underground trolley.

Elec Eng (1894) 17:435

Chester, William S. (? -1900). Electrical engineer.

Eng Rec (1900) 41:209

Chevanne, M. Andre (? -1897). Pioneer mining engineer
on Pacific Coast; inventor in field of hydraulic
mining.

Eng & Min J (1897) 64:582
Eng Rec (1897) 36:509

Chisholm, Henry (1822-81). Iron and steel manufacturer
in Midwest.

AISA Bull (1881) 15:123
Iron Age (5/9/1881) 27:24
Sci Am (1881) 44:374

Christoffel, John B. (1822?-1900). Inventor and manu-
    facturer of boiler-flue cleaner.

    Am Mach (1900) 23:801
    Iron Age (8/9/1900) 66:22

Churchill, A. D. (1856?-96). Mining engineer; author of
    works on mineralogy.

    Eng & Min J (1896) 62:227
    Iron Age (1896) 58:368

Cisneros, Francisco Javier (1836-98). Constructor of
    railroads in Cuba and Colombia.

    ASCE Proc (1898) 24:873-6
    ASCE Trans (1899) 41:622-5
    Eng Rec (1898) 38:179

Clapp, George Mosley (? -1897). Manufacturer of
    machinery.

    Eng Rec (1897) 36:179

Clapp, J. Irving (1856-78). Mining, hydraulic, and civil
    engineer.

    Eng News (1878) 5:162

Clapp, Mertillow R. (1827?-87). Inventor of steam fire
    engine·

    Sanit Eng (1887) 16:270

Clark, Frank B. (? -1899). Engineer and surveyor,
    Nicaragua.

    Eng Rec (1899) 40:563

Clark, Ida Edgar (1852-83). City engineer of Cambridge,
    Mass.

    ASCE Proc (1882) 8:92-3

Clark, Jacob M. (1829-94). Civil engineer, consultant
    for Central Railroad of N.J.

    ASCE Proc (1894) 20:203-4
    Eng Rec (1894-5) 31:75
    RR Gaz (1894) 26:893

Clark, Latimer (1822-98). Electrical engineer, concerned
    with land and submarine telegraphy.

    Sci Am (1898) 79:339

Clark, Nathan B. (? -1892). Naval engineer and inventor.

Eng & Min J (1892) 53:456
Iron Age (1892) 49:826

Clark, Robert M. (1816?-89). Locomotive engineer.

Am Mach (4/4/1889) 12:6

Clark, Robert Neilson (1848-94). Mining engineer.

Am Manuf (1894) 54:442
Eng & Min J (1894) 57:300
Eng Rec (1893-4) 29:280

Clark, William A. (1832?-92). Engineer of Western Union Telegraph Company.

Eng Rec (1891-2) 25:190

Clarke, Benjamin G. (1820-92). Authority on iron and steel.

Am Eng & RR J (1892) 66:432
Am Manuf (1892) 51:324
Eng & Min J (1892) 54:180
Eng Rec (1892) 26:180
Iron Age (1892) 50:337-8*
RR Gaz (1892) 24:624

Clarke, George H. (? -1893). Marine engineer.

Eng (1893) 25:101

Clarke, Henry Wadsworth (1837-92). Civil engineer and surveyor, city engineer of several cities.

Am Eng & RR J (1892) 66:239-40
ASCE Proc (1892) 18:93-4
Eng Rec (1891-2) 25:206

Clarke, L. H. (1830-1900). Chief engineer of Lake Shore & Michigan Railroad.

RR Gaz (1900) 32:208

Clarke, William H. (1812-78). Civil engineer, sewers.

Eng News (1878) 5:249
West Soc Eng Proc (1878-9) 4:118-9

Cleeman, Thomas M. (1843-93). Railroad and bridge engineer in Peru and Chile; author of handbook on railroad engineering practice.

ASCE Proc (1894) 20:69-72
Eng & Min J (1893) 56:598
Eng Rec (1893) 28:406
RR Gaz (1893) 25:878

Clemens, Ernest Victor (1855-93). Mechanical engineer and designer of mining machinery; superintendent of refrigerating machine company.

Am Mach (9/21/1893) 16:11
ASCE Proc (1894) 20:161-2
ASME Trans (1893) 14:1450
Eng & Min J (1893) 56:272
Eng Rec (1893) 28:246
Iron Age (1893) 52:435
Power (10/1893) 13:11
RR Gaz (1893) 25:678

Clements, John Barnwell (1851-97). Civil and railroad engineer.

ASME Trans (1897) 18:1102-3
Assoc Eng Soc J (1897) 18:56-7

Close, Charles S. (1817-79). Surveyor and builder, Philadelphia.

Frank Inst J (1879) 78:358-60.

Close, Walter R. (? -1898). Inventor of shingle machine, water wheel, and fog bell.

Iron Age (6/23/1898) 61:21

Clough, Joel Barber (1823-87). Division engineer of Northern Pacific Railroad.

Am Eng & RR J (1887) 61:448
RR Gaz (1887) 19:580

Clowell, Nathan (? -1893). Builder of railroads in Peru; superintendent for Panama Canal.

Eng Rec (1893) 28:118

Cobb, D. B. (1822?-90). Manufacturer of surface condensers.

Iron Age (1890) 45:262

Cobb, Robert Linah (1840-95). Military engineer; chief engineer of Ohio Southern Railroad.

ASCE Proc (1896) 22:574-5
ASCE Trans (1896) 36:545-6
RR Gaz (1895) 27:396

Cochran, James (1823-94). Pioneer coke manufacturer.

AISA Bull (1894) 28:277
Am Manuf (1894) 55:782
Eng & Min J (1894) 58:539

Codman, Henry Sargent (1864?-93). Landscape engineer for World's Fair in Chicago.

Eng Rec (1892-3) 27:151

Coffin, John (1856-89). Draftsman and designer.

Am Eng & RR J (1889) 63:484
ASME Trans (1890) 11:1151
Eng & Min J (1889) 48:228

Coffin, Zebulon E. (? -1896). Designer and manufacturer of valves used in waterworks systems.

Eng Rec (1896) 34:41

Colahan, John B. (? -1892). U.S. government surveyor; constructor of railroads.

Eng Rec (1891-2) 25:290

Colburn, Ransom H. (? -1897). Engineer; worked on Erie Canal and tunnel in Chicago.

RR Gaz (1897) 29:140

Colburn, Warren (? -1879). Chief engineer of Toledo & Wabash Railroad.

ASCE Proc (1880) 6:4-6
RR Gaz (1879) 11:514

Colburn, Zerah (1832-70). Civil engineer, improvements in freight engines; worked on several engineering journals.

ASCE Proc (1896) 22:97-101
ASCE Trans (1896) 36:546-50
Sci Am (1870) 22:315*

Appleton's Cyc 1:682-3
NYT (5/2/1870) 4:7

Colby, Charles Lewis (1839-96). Civil engineer, shipbuilder, and railroad manager.

AISA Bull (1896) 30:60
Eng Rec (1895-6) 33:219
RR Gaz (1896) 28:170

Coleman, John A. (? -1896). Mechanical engineer.

Eng Rec (1895-6) 33:399

Colin, Alfred (? -1900). Mechanical engineer.

Iron Age (6/21/1900) 65:26

Collin, J. B. (1830?-86). Mechanical engineer of Pennsylvania Railroad.

RR Gaz (1886) 18:236

Collins, Charles (? -1877). Railroad engineer.

RR Gaz (1877) 9:42

Collins, Michael H. (1811?-91). Inventor of quartz
crushing machine.

Eng Rec (1891-2) 25:54

Colman, Isaac D. (1818-75). Civil engineer, railroads
and bridges.

ASCE Proc (1873-5) 1:331-2

Colt, Samuel (1814-62). Inventor and manufacturer of
first practical revolving firearm.

Sci Am (1869) 21:408

Appleton's Cyc 1:694-5*
DAB 4:318-9
NYT (1/11/1862) 5:2

Comstock, Harry (1828?-95). Responsible for introduction
of rice-hulling machines in Far East; inventor of guns.

Am Mach (1895) 18:308

Conant, F. H. (? -1898). Naval engineer.

Am Mach (1898) 21:909

Conant, Thomas P. (1860-91). Mining and electrical engineer.

ASME Trans (1891) 12:1056
Elec Eng (1891) 11:291
Elec World (1891) 17:354
Eng & Min J (1891) 51:265

Congdon, Isaac Hopkins (1833-99). Master mechanic and machinist.

Am Eng & RR J (1899) 73:328
RR Gaz (1899) 31:619

Connor, Addison (1847-91). Railroad and hydraulic engineer; assistant engineer of N.Y. dock department.

ASCE Proc (1896) 22:615-6
ASCE Trans (1896) 36:551
Eng Rec (1890-1) 23:88

Cook, Frank (1864-91). Draftsman and designer of automatic nut tapping machines.

Am Mach (6/25/1891) 14:7

Cook, Frederic (1829-99). Manufacturer of sugar machinery.

ASME Trans (1899) 20:1010-1
Iron Age (3/9/1899) 63:16-7

Cook, George Hammell (1818-89). Civil engineer and geologist; planner for railroads.

Am Eng & RR J (1889) 63:484
AIME Trans (1889) 18:218-22
Iron Age (1889) 44:484
Sci Am (1889) 61:212,265*

Appleton's Cyc 1:714

Cook, Henry F. (1850-1900). Hydraulic engineer, constructor of well plants and wells.

Eng Rec (1900) 41:505
Iron Age (5/24/1900) 65:21

Cook, Osceola (1854?-99). Inventor of hair clipping machinery.

Am Mach (1899) 22:244

Cook, Ransom (1794-1881). Mechanical engineer.

Sci Am (1881) 45:25

Cook, James (1837-83). Locomotive builder.

RR Gaz (1883) 15:530

Cooke, John (1825-82). Manufacturer of cotton machinery and locomotives.

RR Gaz (1882) 14:113
Sci Am (1882) 46:128

NYT (2/19/1882) 7:2

Cooke, Robert L. (1809-77). Civil engineer, railroads and public works.

ASCE Proc (1878) 4:66-7

Cooke, Watts (1833-1900). Master mechanic of Lackawanna Road; president of Passaic Rolling Mill Company.

AISA Bull (1900) 34:173
Am Mach (1900) 23:961
Eng Rec (1900) 42:305
Iron Age (10/4/1900) 66:25
RR Gaz (1900) 32:643,692

Cooke, William (1830-99). Locomotive builder.

Eng & Min J (1899) 68:434
RR Gaz (1899) 31:700

Coon, John (1845-98). Mechanical engineer.

Iron Age (9/1/1898) 62:17

Cooper, John Haldeman (1828-97). Mechanical engineer and machinist.

ASME Trans (1897) 18:1105-6
Eng Rec (1896-7) 35:509
Frank Inst J (1897) 144:154-6*
Mach (1896-7) 3:302

Cooper, Peter (1791-1883). Inventor, designer of first locomotive manufactured in U.S.; manufacturer of structural iron for fireproof buildings.

AISA Bull (1883) 17:101
Am Mach (3/5/1881) 4:8; (4/21/1883) 6:3
Eng & Min J (1883) 35:188
Ind World (7/6/1882) 18:1*,29
Iron Age (4/5/1883) 31:15*
Sci Am (1869) 21:394; (1879) 40:193
Van Nostrand's (1874) 1o:382-3

Appleton's Cyc 1:730-2*
DAB 4:409-10
NYT (4/5/1883) 1:7

Copeland, Charles W. (1815-95). Marine and mechanical engineer; designer of first steam war vessel for U.S. Navy.

Am Eng & RR J (1895) 69:140
ASME Trans (1895) 16:1191
Eng (1890) 19:19*; (1895) 29:48
Eng & Min J (1895) 59:155
Eng Rec (1894-5) 31:183
RR Gaz (1895) 27:113

DAB 4:423

Copeland, George M. (? -1892). Steamboat engineer.

ASME Trans (1892) 13:680

Corbin, H. H. (? -1892). Mining engineer.

Eng & Min J (1892) 54:132

Corliss, George Henry (1817-88). Revolutionized the construction and operation of engines with the Corliss steam engine.

Am Eng & RR J (1888) 62:141
Am Mach (3/10/1888) 11:2; (3/24/1888) 11:5
Am Manuf (3/9/1888) 42:15
Eng & Min J (1888) 45:148
Iron Age (1888) 41:334
Power (3/1888) 8:3*
RR Gaz (1888) 20:129
Sanit Eng (1887-8) 17:204
Sci Am (1888) 58:128,343*

Appleton's Cyc 1:740
DAB 4:441

Cornell, Ezra (1807-74). Helped develop telegraph; supervised in construction of first telegraph line.

Sci Am (1874) 31:401

Appleton's Cyc 1:741
DAB 4:444-6
NYT (12/10/1874) 4:7

Cornell, Thomas Clapp (1819-94). Civil engineer and surveyor; worked on Erie Canal and New York Central Railroad.

Eng Rec (1894-5) 31:93
RR Gaz (1895) 27:28

Corning, Erastus (1794-1872). Iron and steel manufacturer; developer of railroads.

RR Gaz (1872) 4:173
Sci Am (1872) 26:264

Appleton's Cyc 1:742
DAB 4:446-7
NYT (4/10/1872) 8:4

Corscaden, Thomas (1843?-98). Inventor of all wrought steel pulley.

Am Mach (1898) 21:415
Iron Age (5/26/1898) 61:15

Coryell, Martin B. (1815-86). Civil, mining, and hydraulic engineer, canals, railroads, and waterworks.

Am Eng & RR J (1887) 61:10
AIME Trans (1886-7) 15:599-601
ASCE Proc (1889) 15:133-5
RR Gaz (1886) 18:857

Cotter, John (1840-98). Engineer specializing in hydraulic machinery.

Am Mach (1898) 21:927

Cottier, Joseph C. G. (1874-97). Mechanical engineer.

ASME Trans (1898) 19:972

Cottrell, Calvert B. (1821-93). Inventor of improvements in printing press; manufacturer of machinery for textile woodworking.

ASME Trans (1893) 14:1447

DAB 4:462-3

Couch, Alfred B. (1829-88). Mechanical engineer and designer of machine tools.

Am Mach (10/27/1888) 11:7
ASME Trans (1889) 10:832-3
Eng (1888) 16:91

Coulson, Joseph (1861-93). Civil engineer.

Assoc Eng Soc J (1894) 13:134-5

Couzen, Mathew K. (? -1897). Civil engineer.

Eng Rec (1897) 36:443

Cowles, Eugene H. (1854?-92). Discoverer of process of making aluminum with electrical heat.

Elec Eng (1892) 13:465
Elec World (1892) 19:305
Eng Rec (1891-2) 25:358
Frank Inst J (1892) 133:404

Cox, James Bowes (? -1898). Civil engineer of Pa.

Eng Rec (1897-8) 37:315

Cox, Stephen (1822-1900). Manufacturer of heating apparatus.

Am Mach (1900) 23:1123
Iron Age (11/22/1900) 66:44

Coxe, Eckley Brinton (1839-95). Mining engineer, inventor of machinery for coal mining; one of the founders of AIME.

AIME Trans (1895) 25:446
AISA Bull (1895) 29:117
Am Mach (1895) 18:408
Am Manuf (1895) 57:699
ASCE Proc (1896) 22:602
ASCE Trans (1896) 36:552-4
ASME Trans (1895) 16:1182-6*
Cassier's (1892-3) 3:129-32*
Eng (1895) 29:132
Eng Rec (1894-5) 31:435
Ind World (5/30/1895) 44:7
Mach (6/1895-6) 1:6
Power (6/1895) 15:15
RR Gaz (1895) 27:320

Appleton's Cyc 1:761-2
DAB 4:485-6

Craig, Charles A. (1816?-85). Machinist and master mechanic.

RR Gaz (1886) 18:12

Craigin, Henry Adams (1867-96). Mechanical and electrical engineer.

AIEE Trans (1896) 13:443-4

Cramp, Jacob C. (1835?-99). Shipbuilder.

Am Mach (1899) 22:949
Eng Rec (1899) 40:442

Cramp, William (1807-79). Major shipbuilder.

Eng News (1879) 6:226

NYT (7/7/1879) 1:5

Crane, Benjamin F. (1816?-88). Civil engineer, canals
and railroads.

Am Eng & RR J (1888) 62:141
Eng & Min J (1888) 45:57
Sanit Eng (1887-8) 17:124

Craven, Alfred Wingate (1810-79). Civil engineer, chief
engineer of Croton Aqueduct; first president of ASCE.

ASCE Proc (1880) 6:24-6
Eng News (1879) 6:97-8

NYT (3/29/1879) 5:2

Craven, Frank S. (1846-90). Mining engineer.

Eng & Min J (1890) 49:90

Craven, Henry Smith (1845-89). Civil and mining engineer,
construction of Croton Aqueduct.

ASCE Proc (1890) 16:216-7
Eng Rec (1889-90) 21:18

Crawford, Alexander L. (1814-90). Pioneer iron manufacturer of Mahoning Valley.

Am Eng & RR J (1890) 64:238
AISA Bull (1890) 24:99
Am Manuf (4/11/1890) 46:13
Eng & Min J (1890) 49:365
Iron Age (1890) 45:558
RR Gaz (1890) 22:257

Creamer, William G. (1826?-98). Inventor of Creamer car
brake, a type of air brake.

Am Eng & RR J (1898) 72:168
Am Mach (1898) 21:317
Iron Age (4/28/1898) 61:16
RR Gaz (1898) 30:316

Cregier, DeWitt C. (1829-98). City engineer of Chicago;
president of Western Society of Engineers.

Eng Rec (1898) 38:509
Iron Age (11/17/1898) 62:22
RR Gaz (1898) 30:837
West Soc Eng J (1899) 4:122-3*

Crehore, John D. (1826-84). Civil engineer, author of major work on bridge building.

Assoc Eng Soc J (1884-5) 4:81-4

Crichton, Alexander F. (? -1898). Mechanical engineer, improver of jute machinery.

Eng Rec (1897-8) 37:435

Crittenden, William L. (1845?-90). Civil engineer and government surveyor.

Eng Rec (1890) 22:2

Crocker, Charles (1822-88). One of the organizers and builders of transcontinental railroad.

Eng & Min J (1888) 46:133
RR Gaz (1888) 20:548

Appleton's Cyc 2:11
DAB 4:552
NYT (4/15/1888) 5:3

Crocker, Frederick (1831?-95). Pioneer oilman.

Eng & Min J (1895) 59:203

Crocker, Samuel L (1805?-83). Manufacturer of locomotives.

RR Gaz (1883) 15:115

NYT (2/11/1883) 7:3

Crofts, John J. (? -1897). Marine engineer of West Coast.

Eng Rec (1897-8) 37:91

Cronise, Ernest Stoll (1861-96). Mechanical and consulting engineer; railway expert.

ASME Trans (1897) 18:1093
Eng Rec (1896) 34:325
RR Gaz (1896) 28:692

Cronk, Charles W. (? -1899). Engineer, marine construction.

Am Mach (1899) 22:177

Crosby, C. O. (? -1880). Inventor of machines for making ruffles, pointed tape trimming, fish hooks, pins, etc.

Sci Am (1880) 43:370

Crouthers, James A. (1846-91). Naval engineer.

ASME Trans (1891) 12:1057

Crowley, Charles Lee (1863?-97). Electrician.

Elec Eng (1897) 23:649
Elec World (1897) 29:747

Cruso, Frederick D. (? -1895). Civil engineer.

Eng Rec (1895) 32:363

Cullum, George W. (1809-92). Military engineer, author
of several engineering works.

Am Eng & RR J (1892) 66:193
Eng Rec (1891-2) 25:222
RR Gaz (1892) 24:182

Appleton's Cyc 2:27
NYT (6/29/1892) 5:4

Cummer, Franklin D. (1843?-98). Manufacturer of
machinery, inventor of Cummer drier; authority on
mechanical evaporation of garbage.

Eng & Min J (1898) 65:648
Iron Age (5/19/1898) 61:17

Cummings, Benjamin (1772-1843). Inventor of circular saw.

Iron Age (2/2/1882) 29:30

Cunningham, James (1855-97). Mining engineer.

Eng & Min J (1897) 63:95

Curtis, Robert (1835-87). Master mechanic.

Am Eng & RR J (1888) 62:45
RR Gaz (1887) 19:819

Curtis, Wendell R. (1850-93). Civil engineer, specialist
in river and harbor work.

ASCE Proc (1894) 20:86

Curtiss, G. F. (? -1895). Electrical engineer.

Elec Eng (1895) 20:451

Curtiss, William Giddings (? -1900). Railroad engineer.

RR Gaz (1900) 32:433

Cushing, D. L. (? -1895). Civil engineer.

Eng Rec (1895) 32:363

Cushing, George (1816-87). Marine engineer.

Eng (1887) 13:89

Cushing, Oliver E. (1829-90). Civil engineer and
mechanical draftsman of Lowell Gaslight Company.

ASCE Proc (1890) 16:166

Cushing, Samuel B. (1811-73). Civil and hydraulic engineer, railroads and bridges.

ASCE Proc (1873-5) 1:43-4
RR Gaz (1873) 5:311
NYT (7/18/1873) 5:4

Dabney, Frank Yeamans (1835-1900). Railroad engineer.

Eng Rec (1900) 41:330
RR Gaz (1900) 32:227

Daddow, Samuel Harries (1827?-75). Miner and inventor.

Eng & Min J (1875) 19:225

Dagron, James G. (? -1895). Constructor of bridges for
Baltimore & Ohio Railroad; electric street railroads
in Philadelphia.

ASME Trans (1896) 17:738
RR Gaz (1895) 27:354

Dahlgren, John Adolphus Bernard (1809-70). Military
engineer, inventor of naval guns and ordnance.

Sci Am (1870) 23:58

Appleton's Cyc 2:53-4*
DAB 5:29-31
NYT (7/15/1870) 5:2

Dale, Samuel (1820?-96). Railroad surveyor and engineer
in Eastern U.S.

Eng & Min J (1896) 61:523
Eng Rec (1895-6) 33:453

Danforth, Charles (1797-1876). Inventor of Danforth spindle and cap-spinning frame; locomotive builder.

RR Gaz (1876) 8:135,144

Appleton's Cyc 2:72-3
DAB 5:65-6
NYT (3/23/1876) 2:4

Danforth, William (? -1897). Civil engineer, Minn.

Eng Rec (1897-8) 37:3

Daniels, Henry A. (? -1899). Chief engineer of Sing Sing water works.

Eng Rec (1899) 40:110
Iron Age (6/29/1899) 63:21

Darling, Benjamin (1808?-90). Inventor of revolving-type pistol.

Iron Age (1890) 45:647

Darling, Edwin (? -1898). Water works commissioner of Pawtucket, R.I.

Eng Rec (1898) 38:399

Darling, Robert F. (1868?-99). Draftsman and mechanic.

Am Mach (1899) 22:158

Darling, Samuel (1815-96). Inventor of improvements in saw mill machinery; manufacturer of machinist tools.

Iron Age (1896) 58:779

Darracott, Franklin (1820-95). Civil engineer.

Eng Rec (1894-5) 31:165

Dart, Henry (? -1886). Civil engineer, bridges and railroads.

RR Gaz (1886) 18:117

Dashiell, Robert B. (1861-99). Assistant naval constructor; inventor of rapid fire breech mechanism.

Iron Age (3/16/1899) 63:18
RR Gaz (1899) 31:216

Davenport, Thomas (1802-51). Inventor of electric motor.

Elec World (1891) 17:228-9

Appleton's Cyc 2:84-5
DAB 5:87-8

Davenport, William R. (1831?-88). Manufacturer of car wheels and pig iron.

Am Eng & RR J (1889) 63:47
AISA Bull (1888) 22:370
Iron Age (1888) 42:938
RR Gaz (1888) 20:843

David, Emil (? -1883). Civil engineer.

RR Gaz (1883) 15:65

Davidson, George (? -1895). Designer and engineer for refrigerating plants.

ASME Trans (1896) 17:739

Davidson, Matthias O. (1819-72). Civil engineer, railroads in Cuba and mines in Arizona.

ASCE Proc (1893) 19:56-7
RR Gaz (1872) 4:396

Davies, Frederick (? -1887). Electrician; superintendent of South American Telegraph Company.

Sanit Eng (1886-7) 15:142

Davis, Augustus B. (1815?-91). Inventor of instruments of precision and railroad car springs; manufacturer of weighing machinery.

Iron Age (1891) 47:785

Davis, Daniel (1813-87). Electrician and electro-mechanician.

Elec Eng (1887) 6:165
Elec World (1887) 9:166

Davis, David T. (1834?-85). Master mechanic, machinist, and superintendent of shops.

Nat Car (1885) 16:98

Davis, Edward F. C. (1847-95). Mechanical engineer, president of ASME.

Am Eng & RR J (1895) 69:429-30*,474
ASME Trans (1895) 16:1177-81*
Eng (1895) 30:43
Eng & Min J (1895) 60:131
Eng News (1895) 33:69*; (1895) 34:89
Eng Rec (1895) 32:183
Iron Age (1895) 56:335*
Mach (9/1895-6) 2:21*
RR Gaz (1895) 27:536,543*
Sci Am (1895) 73:134

54

Davis, Frank P. (? -1900). Railway engineer in South
    America.

Eng Rec (1900) 41:454

Davis, James (? -1897). Civil engineer; worked on
    Croton Aqueduct and Mohawk Canal.

Eng Rec (1897) 36:289
Iron Age (9/9/1897) 60:13

Davis, Robert L. (1823?-92). Civil engineer.

Eng Rec (1892) 26:150

Davis, Thomas W. (? -1893). City surveyor of Boston.

Eng Rec (1892-3) 27:433,453

Davison, Henry J. (1835-90). Mechanical and gas engineer;
    iron structural work.

Am Eng & RR J (1890) 64:428
Am Mach (7/31/1890) 13:8
ASME Trans (1890) 11:1154-5
Eng & Min J (1890) 50:106
Eng Rec (1890) 22:114
Iron Age (1890) 46:181

Dawson, Walter (1823-98). Master mechanic of Delaware,
    Lackawanna & Western Railroad; inventor of miner's
    lamp.

Eng & Min J (1898) 65:468
RR Gaz (1898) 30:284

Day, Austin Goodyear (1824-90). Inventor in the field of
    rubber vulcanization and telegraph cables.

Elec Eng (1890) 9:75
Elec World (1890) 15:82

Dean, George W. (? -1897). Member of engineering and
    astronomical division of U.S. Coast Survey.

Eng Rec (1896-7) 35:179

Dean, Ward Hunt (1860?-1900). Inventor and manufacturer
    of pumps.

Am Mach (1900) 23:47
Am Manuf (1900) 66:36
Iron Age (1/11/1900) 65:29

Dearborn, William Lee (1812-75). Civil engineer, rail-roads; improvements in N.Y. City public works.

ASCE Proc (1873-5) 1:330-1

Debes, J. C. (1835-98). Mechanical engineer.

ASME Trans (1899) 20:1005-6

Debuol, Ulysses G. Scheller (1816-95). Civil engineer.

Eng & Min J (1895) 59:59
Eng Rec (1894-5) 31:129

DeCamp, Alfred H. (1856?-95). Inventor and mechanical engineer.

Eng & Min J (1895) 60:12
Eng Rec (1895) 32:93
Iron Age (1895) 56:23

Dechert, William Wirt (1834-79). Civil engineer, builder of railroads in Cuba.

RR Gaz (1897) 11:556

DeCoudres, Louis (1789-1872). Helped fabricate the machinery of the first steamboats.

Sci Am (1873) 28:33

Appleton's Cyc 2:122-3

DeCourcy, Bolton W. (? -1900). Railway engineer.

Eng Rec (1900) 41:354

Deegan, Thomas (? -1897). Mechanical engineer.

Eng Rec (1897) 36:443

Deere, John (1804-86). Manufacturer of first steel plow in U.S.

Ind World (5/22/1886) 26:6-7

DAB 5:193-4

Degenhardt, Frederick (1856-94). Inventor of electrical devices.

Elec Eng (1894) 18:487

DeGurse, Joseph (1857-98). Chief engineer of Lake Erie & Detroit River Railroad.

Assoc Eng Soc J (1898) 20:62

Degnon, John (1810?-69). Locomotive engineer, first man to drive a locomotive in the U.S.

Sci Am (1869) 21:409

Degraff, Alonzo H. (? -1893). Civil engineer, railroads in N.Y. State.

Eng Rec (1893) 28:262

DeHart, Norwood (1872-99). Engineer of Chicago Drainage Canal; elevated railroads in Chicago and Boston.

West Soc Eng J (1899) 4:356

DeKalb, Enoch E. (? -1899). Inventor in the field of car ventilation.

RR Gaz (1899) 31:270

DeLacy, Walter Washington (1819-92). Civil engineer, railroads.

Assoc Eng Soc J (1897) 18:341-50*
Eng Rec (1892) 26:2

DAB 5:206-7

Delamater, Cornelius H. (1821-89). Mechanical engineer; builder of first iron boats and steam fire engines in U.S.

Am Eng & RR J (1889) 63:145
AISA Bull (1889) 23:45
Am Mach (2/14/1889) 12:7
ASME Trans (1889) 10:836-8
Eng (1889) 17:43
Eng & Min J (1889) 47:146
Eng Rec (1888-9) 19:152
Iron Age (1889) 43:248
RR Gaz (1889) 21:118
Sci Am (1889) 60:97

DAB 5:211-2

Delano, Benjamin Franklin (1809-82). Naval constructor.

Sci Am (1882) 46:298
NYT (5/1/1882) 5:1

De la Vergne, John C. (1840-96). Manufacturer and inventor of refrigerating and ice making machinery for breweries.

Am Mach (1896) 19:523
Eng & Min J (1896) 61:475
Eng Rec (1895-6) 33:417
Sci Am (1896) 74:359

DeLuse, Edmund S. (1829-90). Chief engineer of U.S. Navy.

Iron Age (1890) 46:13

Deming, John (1817-94). Manufacturer of pumps and hydraulic machinery.

Iron Age (1894) 53:166

Demmler, J. H. (1808?-93). Founder of U.S. Iron & Tin Plate Manufacturing Company.

Am Manuf (1894) 54:356
Iron Age (1893) 52:391

Denison, Charles H. (? -1898). Inventor of woodworking machinery and water filters.

Eng Rec (1898-9) 39:61

Dennett, Joseph G. (? -1896). Mechanical engineer; chief engineer of waterworks at Salem, Mass.

Eng (1896) 32:125*
Eng Rec (1896) 34:421

Dennison, Albert P. (? -1886). Chief engineer of Merritt Coast & Wrecking Company; docks at Staten Island.

Sanit Eng (1885-6) 13:280

Dent, Edward L. (1861-99). Manufacturer of structural iron work.

ASME Trans (1900) 21:1160

DePauw, Washington Charles (1822-87). Manufacturer of plate glass.

Am Eng & RR J (1887) 61:247
AISA Bull (1887) 21:124
Iron Age (5/12/1887) 39:17

Appleton's Cyc 2:144
DAB 5:244

Detmold, Christian E. (1810-87). Civil and mining
    engineer; supervising architect and engineer for
    Crystal Palace, N.Y. City.

    Am Eng & RR J (1887) 61:349
    AISA Bull (1887) 21:177
    Eng & Min J (1887) 44:19
    Sanit Eng (1887) 16:158

    Appleton's Cyc 2:154
    DAB 5:258

Devereux, John Henry (1832-86). Civil engineer, rail-
    roads.

    RR Gaz (1886) 18:220,225*

    Appleton's Cyc 2:156
    DAB 5:262-3

Dewitt, David P. (1817-89). Civil engineer.

    Am Eng & RR J (1889) 63:196

Dialogue, John H. (1828-98). Shipbuilder.

    Am Eng & RR J (1898) 72:418
    Am Mach (1898) 21:811
    AISA Bull (1898) 32:172
    Eng & Min J (1898) 66:524
    Eng Rec (1898) 38:465
    Iron Age (10/27/1898) 62:21

Dick, Robert (1814-90). Inventor of newspaper mailing
    machine.

    Sci Am (1890) 63:385

    Appleton's Cyc 2:169-70

Dickerson, Edward N. (1824?-89). Mechanical engineer
    and inventor; noted patent lawyer.

    Elec Eng (1890) 9:33
    Elec World (1889) 14:406

    NYT (12/13/1889) 5:5

Dickinson, Charles Wesley (1823?-1900). Inventor of
    lathe used in engraving.

    Am Mach (1900) 23:675

Dickinson, George Codwise (1832-92). Mining and rail-
    road engineer.

    Am Eng & RR J (1892) 66:146
    ASCE Proc (1892) 18:71-2
    RR Gaz (1892) 24:89

Dickinson, Pomeroy P. (1827?-95). Civil engineer, railroads and bridges in N.Y. State.

ASCE Proc (1899) 25:1057-8
Eng & Min J (1895) 60:351
Eng Rec (1895) 32:345
RR Gaz (1895) 27:679

Dickson, Thomas (1822-84). Manufacturer of steam engines
and mining machinery.

AISA Bull (1884) 18:205
Iron Age (8/7/1884) 34:21
RR Gaz (1884) 16:593
Sci Am (1884) 51:81

Appleton's Cyc 2:176
DAB 5:306
NYT (4/1/1884) 5:3

Dietz, Robert Edwin (1807?-97). Pioneer manufacturer of
gas fixtures.

Sci Am (1897) 77:219

Dillman, G. N. (? -1898). Civil engineer.

Eng Rec (1898) 38:25

Dillon, Sidney (1812-92). Railroad contractor and
builder of Union Pacific Railroad.

Am Eng & RR J (1892) 66:337
ASCE Trans (1896) 36:603-4
Nat Car (1892) 23:110
RR Gaz (1892) 24:455

DAB 5:312
NYT (6/10/1892) 9:3

Disston, Henry (1819-78). Inventor of Disston saw;
manufacturer of cutting tools and steel implements.

Ind World (1/14/1886) 26:22
Iron Age (3/21/1878) 21:1*

Appleton's Cyc 2:182
DAB 5:318-9

Disston, Thomas S. (1833-97). Inventor and manufacturer.

AIME Trans (1899) 29:xviii
Eng & Min J (1897) 64:672
Iron Age (11/25/1897) 60:18

Dixon, Joseph (1799-1869). Inventor and manufacturer of lithographic equipment.

Sci Am (1869) 21:6

Appleton's Cyc 2:186-7
DAB 5:329-30

Doane, Thomas (1821-97). Civil and mechanical engineer; chief engineer of Hoosac Tunnel; pioneer in use of compressed air machinery.

ASCE Trans (1898) 39:690-4
Eng & Min J (1897) 64:522
Eng Rec (1897) 36:465
Iron Age (10/28/1897) 60:18
RR Gaz (1897) 29:773-4
Sci Am (1897) 77:295

DAB 5:334

Dodge, Miles Benjamin (1829-96). Inventor of mining machinery.

Eng & Min J (1896) 62:299

Dodge, Wallace H. (1849?-94). Inventor of continuous system of rope drive pulleys.

Am Mach (9/27/1894) 17:12
ASME Trans (1894) 15:1192-3
Power (10/1894) 14:15

Doerflinger, Augustus (? -1899). Government engineer; discoverer of method of exploding blasting charges.

Eng Rec (1899) 40:635

Doig, William S. (1848?-1900). Inventor and manufacturer of box nailing machinery.

Am Mach (1900) 23:305

Donkin, Bryan (? -1893). Mechanical engineer; manufacturer of mill machinery and steam engines.

RR Gaz (1893) 25:936

Dopson, R. W. (? -1899). Civil engineer, street railways in Savannah.

Eng Rec (1899) 40:635

Doran, Frank C. (1849-98). Railway engineer, elevated railways in Chicago.

Eng Rec (1898) 38:531
RR Gaz (1898) 30:767

Doran, James Shreve (1834?-1900). Superintendent and constructing engineer of International Navigation Company.

Am Mach (1900) 23:1245

Dorr, Frank Hayes (1869-97). Electrical engineer.

AIEE Trans (1897) 14:615-6

Dougherty, John (1803?-86). Inventor of portable iron section boats.

RR Gaz (1886) 18:822

Doughty, Samuel Sitwell (1810?-88). Surveyor of Central Park, N.Y. City.

Am Eng & RR J (1888) 62:381
Eng Rec (1888) 18:82

Douglas, Benjamin (1816-94). Inventor and manufacturer of revolving cistern stand pump.

Iron Age (1894) 54:16,79*

DAB 5:394-5

Douglass, David Bates (1790-1849). Civil engineer, water works; U.S. Corps of Engineers.

Van Nostrand's (1872) 6:1-6*

Appleton's Cyc 2:216-7
DAB 5:405-6

Douglass, William G. (1819?-91). Civil engineer, railroads.

Am Eng & RR J (1891) 65:187
Eng Rec (1890-1) 23:254
Nat Car (1891) 22:58

Dow, Lorenzo (1825-99). Inventor of type setting machine and waterproof cartridge.

Eng & Min J (1899) 68:494
Eng Rec (1899) 40:491

DAB 5:410-1

Dowd, Thomas F. (? -1898). Marine engineer.

Eng (1898) 35:143

Downes, Stancliff Bazen (1859-95). Civil and topographical engineer.

ASCE Proc (1895) 21:181-2
Eng & Min J (1895) 59:395
Eng Rec (1894-5) 31:381

Drake, Alexander E. (? -1897). Topographical engineer.

Eng Rec (1896-7) 35:333

Drake, E. F. (1814?-92). Railroad engineer in West.

Eng Rec (1891-2) 25:190
RR Gaz (1892) 24:145

Drake, E. L. (? -1880). First man to sink an oil well in U.S.

AISA Bull (1880) 14:282
Sci Am (1880) 43:344

NYT (11/10/1880) 5:3

Drake, James S. (? -1896). Master mechanic, superintendent of New Jersey & New York Railroad.

RR Gaz (1896) 28:296

Dresser, George W. (? -1883). Civil engineer, railroads and public works; editor of American Gas Light Journal.

ASCE Proc (1893) 19:110
Sanit Eng (1883) 8:16

Duane, James Chatham (1824-97). Mining and military engineer; author of manual for engineer troops.

ASCE Trans (1898) 39:686-90
Eng Rec (1897) 36:509
RR Gaz (1897) 29:805; (1899) 31:51
Sanit Eng (1886) 14:501

Appleton's Cyc 2:236
DAB 5:466-7
NYT (11/9/1897) 7:6

DuBarry, Joseph Napoleon (1830-92). Civil engineer, railroads.

Am Eng & RR J (1893) 67:50
ASCE Proc (1893) 19:108-9
Eng Rec (1892-3) 27:70
Nat Car (1893) 24:10
RR Gaz (1892) 24:974

Dudgeon, Richard (1819?-95). Inventor of hydraulic jack.

AISA Bull (1895) 29:93
Eng Rec (1894-5) 31:345
RR Gaz (1895) 27:240

Duer, Augustus P. (? -1898). Civil engineer.

Eng Rec (1897-8) 37:381

Duff, Thomas (1851-98). Marine engineer, iron works proprietor.

Am Mach (1898) 21:625
Iron Age (8/11/1898) 62:18

Dunbar, Oliver P. (? -1897). Superintendent of Motive Power and Cars of Wheeling & Lake Erie Railroad.

RR Gaz (1897) 29:805

Dunbar, Robert (1812?-90). Mechanical engineer, inventor of grain elevators; organized Eagle Iron Works.

Eng & Min J (1890) 50:368
Eng Rec (1890) 22:258
RR Gaz (1890) 22:671

DAB 5:506-7

Duncomb, Jesse S. (? -1885). Engineer.

Eng (1885) 10:5

Dunham, John (1820-88). Mechanical engineer, builder of ironclads and the piers of the Harlem Bridge.

Iron Age (1888) 41:190

Dunn, B. R. (? -1893). Railroad engineer.

RR Gaz (1893) 25:629

Dunne, Henry W. (? -1894). Civil engineer, superintendent of New York, Philadelphia, & Norfolk Railroad.

RR Gaz (1894) 26:603

Dunning, William B. (1818-89). Machinist, inventor of steam heating boilers.

Eng Rec (1896) 34:41
Sci Am (1889) 61:243*

DuPont, Henry (1812-89). Manufacturer of explosives.

Iron Age (1889) 44:250
RR Gaz (1889) 21:546
DAB 5:528

DuPuy, T. Haskins (? -1890). Civil engineer, railroads.

Am Eng & RR J (1890) 64:281
RR Gaz (1890) 22:384

Durfee, William Franklin (1833-99). Engineer, inventor, and architect; supervised the making of the first Bessemer steel produced in U.S.

AIME Trans (1900) 30:xxix-xxx
AISA Bull (1899) 33:302
Am Mach (1899) 22:1118-9
ASME Trans (1900) 21:1161-2
Eng & Min J (1899) 68:615
Eng Rec (1899) 40:586
Iron Age (11/23/1899) 64:19
RR Gaz (1899) 31:799,818

Appleton's Cyc 2:271-2
DAB 5:547-8

Durham, Leroy T. (? -1896). Civil engineer.

Eng Rec (1896-7) 35:25

Dutro, T. C. (? -1886). Manufacturer of car wheels.

RR Gaz (1886) 18:170

Dwight, George S. (? -1886). Gas engineer.

Eng & Min J (1886) 42:171

Dwight, William Smith (1861-83). Civil engineer; inspector of iron bridge work.

Am Manuf (6/8/1883) 32:13

Dyer, Gustave (? -1898). Engineer under James B. Eads.

Eng Rec (1898) 38:421

Dyer, Paul D. (1860?-90). Electric light engineer.

Elec World (1890) 15:66

Eads, James Buchanan (1820-87). Designer of steel arch
   bridge at St Louis; responsible for jetty works at
   mouth of Mississippi.

   Am Eng & RR J (1887) 61:150-1
   Am Mach (4/2/1887) 10:3
   ASCE Proc (1887) 13:46-65
   Assoc Eng Soc J (1890) 9:456-61
   Cassier's (1893-4) 5:233-6*
   Eng News (1887) 17:174,256*
   Ind World (3/24/1887) 28:7
   Iron Age (3/17/1887) 39:21
   Pop Sci (1886) 28:544*; (1887) 31:144
   RR Gaz (1887) 19:176*
   Sanit Eng (1886-7) 15:406
   Sci Am (1876) 34:81; (1887) 56:176,263-4*

   Appleton's Cyc 2:287*
   DAB 5:587-9
   NYT (3/11/1887) 5:4

Eames, Frederick W. (? -1883). Inventor of Eames
   vacuum brake.

   RR Gaz (1883) 15:271

East, Rowland H. (? -1900). Draftsman and railway
   engineer.

   Iron Age (2/1/1900) 65:21

Eastwick, Andrew M (1810?-79). Inventor and locomotive
   designer.

   RR Gaz (1879) 11:88

Eastwood, Benjamin (1839?-99). Founder of Eastwood Machine Works, specializing in laundry machinery.

   Am Mach (1899) 22:419
   Iron Age (5/4/1899) 63:18

Eaton, Horace LaFayette (1851-95). Civil engineer,
   public works.

   ASCE Proc (1896) 22:616-7
   ASCE Trans (1896) 36:554-5
   Assoc Eng Soc J (1897) 19:25
   Eng News (1895) 34:376; (1896) 35:24
   Eng Rec (1895) 32:471

Eaton, Luther H. (1821-78). Civil engineer, railroads
   and dams.

   Eng News (1878) 5:42

Eaton, S. D. (1823-99). Civil engineer of Iowa.

Eng Rec (1899) 40:278

Eayrs, Norman W. (? -1900). Railway engineer.

Eng Rec (1900) 41:481

Eckert, Edward (? -1890). Inventor of telephonic apparatus.

Elec Eng (1896) 22:139*

Eddy, George W. (1810-97). Inventor and manufacturer of Eddy valve and fire hydrant.

Am Mach (1897) 20:967
Iron Age (12/16/1897) 60:17

Eddy, Luther Devotion (1811?-92). Surveyor, N.Y. State.

Eng Rec (1892) 26:226

Eddy, Wilson (1813-98). Locomotive engineer, master mechanic of Boston & Albany Railroad.

Eng Rec (1898) 38:311
RR Gaz (1898) 30:655

Edes, Oliver (1816?-84). Inventor of rivet machines.

Iron Age (2/28/1884) 33:30

Edge, George W. (1811-80). Gas engineer.

ASCE Proc (1880) 6:18-9

Edgerton, Benjamin Hyde (1811-86). Civil engineer, railroads.

RR Gaz (1886) 18:894

Edmonds, William H. (1865?-98). Mining engineer; founder of Manufacturer's Record.

AISA Bull (1898) 32:61
Eng & Min J (1898) 66:284
Iron Age (4/7/1898) 61:18

Edson, Marmont B. (1813-92). Inventor of pressure recording gauge.

Elec World (1892) 19:271
Sci Am (1892) 66:374

Edwards, John (1801?-79). Locomotive builder.

RR Gaz (1879) 11:216

Eekhout, Bernhard (1868?-97). Electrical engineer.

Elec World (1897) 29:492

Egleston, Thomas (1832-1900). Founder of School of Mines at Columbia University; professor of metallurgy and mineralogy.

AIME Trans (1901) 31:3
AISA Bull (1900) 34:18
Am Manuf (1900) 66:47
Eng & Min J (1897) 64:5*; (1900) 69:74
Iron Age (1/18/1900) 65:33
RR Gaz (1900) 32:46

Appleton's Cyc 2:315-6
DAB 6:56

Eickemeyer, Rudolph (1831-95). Inventor and manufacturer of Eickemeyer dynamo and other mechanical and electrical devices.

AIEE Trans (1894) 11:876-7; (1895) 12:668
Elec Eng (1890) 10:669-70*; (1895) 19:106,133-4*
Elec World (1895) 25:331-2*

DAB 6:59-60

Einsiedel, D. (? -1896). Civil and mechanical engineer of New Orleans.

Eng Rec (1896) 34:325

Eldredge, C. J. (? -1892). Civil and mining engineer.

Eng & Min J (1892),54:468

Elliot, George Hartford (1842-86). Military engineer.

ASCE Proc (1889) 15:110-1

Elliott, George Henry (1831-1900). Military engineer, fortifications and lighthouses.

Eng Rec (1900) 41:306

Appleton's Cyc 2:328

Ellis, Charles (? -1898). City engineer of Kansas City.

Eng Rec (1898) 38:135

Ellis, Charles G. (1843?-91). Manufacturer of locomotives.

Iron Age (1891) 47:978
Nat Car (1891) 22:97*

Ellis, Charles H. (? -1894). Civil engineer.

Eng Rec (1894-5) 31:21

Ellis, James L. (1841?-99). Mechanic of Boston & Maine Railroad.

Am Mach (1899) 22:77

Ellis, Mathew (1839-92). Master mechanic.

Nat Car (1893) 24:10

Ellis, Nathaniel Webster (1849-89). Civil engineer, railroads; water works in New England.

ASCE Proc (1889) 15:112-3

Ellis, Theodore G. (1829-83). Military and civil engineer; inventor of current meter; conducted several hydraulic experiments.

ASCE Proc (1897) 23:324-5
ASCE Trans (1897) 37:557-9
Eng News (1883) 10:19
RR Gaz (1883) 15:31

Appleton's Cyc 2:334
NYT (1/10/1883) 5:4

Ellmaker, Frank (1844?-99). Railroad engineer.

RR Gaz (1899) 31:852

Ellsworth, A. B. (1865-93). Civil engineer, surveyor, and bridge inspector.

ASCE Proc (1893) 19:48
Eng Rec (1892-3) 27:151

Elseffer, William L. (1828?-98). Civil engineer, railways and Central Park; responsible for drainage of salt meadows in N.J.

Eng & Min J (1898) 65:48
Eng Rec (1897-8) 37:117
Iron Age (1/6/1898) 61:21
RR Gaz (1898) 30:16

Ely, Amasa (? -1896). Civil engineer, Bureau of Water, Philadelphia.

Eng Rec (1896-7) 35:25

Ely, George H. (1825-94). Iron manufacturer; president of Western Iron Ore Association.

AISA Bull (1894) 28:29
Eng & Min J (1894) 57:108
Iron Age (1894) 53:230

Emack, Charles S. (1834-77). Civil engineer, railroads.

ASCE Proc (1878) 4:65

Embick, Frederick (? -1891). Railroad engineer.

Eng Rec (1891) 24:150

Emerson, B. F. (1837-87). Superintendent of Copper Falls Mining Company, Michigan.

ASME Trans (1887) 8:727-8

Emerson, James E. (? -1896). Inventor of continuous steam heating system.

RR Gaz (1896) 28:497

Emerson, James E. (1823-1900). Inventor of inserted tooth for circular saws.

Am Manuf (1900) 66:575*

Appleton's Cyc 2:342
DAB 6:129

Emery, Charles Edward (1838-98). Electrical engineer and builder of steam engines; authority on distribution of heat and power from a central steam plant.

Am Eng & RR J (1898) 72:234
AIEE Trans (1898) 15:737-43
AIME Trans (1899) 29:viii-xxix
Am Mach (1898) 21:435,454
ASCE Proc (1899) 25:738-40
ASCE Trans (1899) 42:558-60
ASME Trans (1898) 19:977-9
Cassier's (1893) 4:103-6
Elec Eng (1898) 25:653
Elec World (1898) 31:730
Eng (1890) 20:43-4*
Eng & Min J (1898) 65:678
Eng Rec (1898) 38:3
Iron Age (6/9/1898) 61:19*
Power (7/1898) 18:19
RR Gaz (1898) 30:419

Appleton's Cyc 2:349
DAB 6:142-3

Emery, Horace L. (? -1892). Manufacturer and improver of agricultural implements and machinery.

Iron Age (1892) 49:415

Emmet, Thomas Addis (1818-80). Civil engineer and surveyor, Erie & Hudson River Railroad.

ASCE Proc (1880) 6:2-4

Appleton's Cyc 2:350

Emonts, William Alexis George (1847-87). Railroad engineer and draftsman.

ASCE Proc (1896) 22:578
ASCE Trans (1896) 36:594

Emory, William Hemsley (1811-87). Military and topographical engineer.

Am Eng & RR J (1888) 62:46
Sanit Eng (1887-8) 17:28

Appleton's Cyc 2:352
DAB 6:153-4
NYT (12/3/1887) 4:7

Emrich, Anton F. (1859-93). Mining engineer and metallurgist.

Eng & Min J (1893) 56:526

Emslie, Peter (? -1887). Civil engineer.

Am Eng & RR J (1887) 61:246

Endlich, Frederick M. (1851-99). Mining engineer and metallurgist of Calif.

Eng & Min J (1899) 68:126*

Endres, John J. (1842?-89). Chief engineer of Hudson County Elevated Cable Railroad, Hoboken, N.J.

Eng Rec (1889) 20:24

Englemann, Henry (1831-99) Mining engineer and metallurgist; consulting engineer.

AIME Trans (1900) 30:xxx

Ennis, William (1822-81). Inventor of improvements in furnaces.

Sci Am (1881) 44:308

Ensley, Enoch (1836-91). Pioneer in development of iron industry in South.

Iron Age (1891) 48:985

DAB 6:169

Ericsson, John (1803-89). Civil and mechanical engineer, designer of screw propeller and of first armored turret ship, the Monitor.

Am Eng & RR J (1889) 63:151-6*
Am Mach (3/14/1889) 12:8
Am Manuf (3/15/1889) 44:11
ASME Trans (1889) 10:845-50
Appleton's Mech Mag (1852) 2:259-61
Cassier's (1894-5) 7:3-21*,163-76*,215-27
Eng (1889) 17:66,78-9
Eng & Min J (1889) 47:229,234
Eng News (1881) 8:403; (1889) 21:339-40*
Eng Rec (1888-9) 19:194
Ind World (8/9/1888) 31:10; (3/14/1889) 32:1
Iron Age (1889) 43:402-3
Pop Sci (1889) 35:144; (1893-4) 44:112-20*
Power (4/1889) 9:13
RR Gaz (1889) 21:186
Sci Am (1862) 6:209-10*; (1869) 21:409
        (1889) 60:163-4*,175,182-3; (1890) 63:128-9*

Appleton's Cyc 2:363-5*
DAB 6:171-6
NYT (3/9/1889) 1:7

Essick, Samuel V. (1840?-1900). Inventor of printing telegraph.

Elec World (1900) 36:557

Eunson, Robert Groat (1806-96). Marine engineer, responsible for improvements in marine engines.

Eng & Min J (1896) 61:547
Eng Rec (1896) 34:3
Iron Age (1896) 57:1316

Evans, Louis Provost (? -1896). Civil engineer and bridge builder.

Eng Rec (1896) 34:217

Evans, Anthony Walton White (1817-86). Civil engineer,
    railroads in Chile and New Zealand; responsible for
    first steam railroad in South America.

    Am Eng & RR J (1887) 61:9-10
    ASCE Proc (1887) 13:30-5
    RR Gaz (1886) 18:827,834
    Sanit Eng (1886-7) 15:21

Everson, William H. (1817-96). "Father" of sheet iron
    industry.

    AISA Bull (1896) 30:92
    Am Manuf (1896) 58:551
    Eng & Min J (1896) 61:379
    Iron Age (1896) 57:926

Ewbank, Thomas (1792-1870). Inventor of improved methods
    of tinning lead and improved steam safety valves;
    manufacturer of copper, lead, and tin tubing.

    Nat Car (11/1870-1) 1:7
    Sci Am (1870) 23:213

    Appleton's Cyc 2:391
    DAB 6:227-8

Ewen, John (1810?-77). Civil engineer; president of
    Pennsylvania Coal Company.

    Eng & Min J (1877) 23:351
    RR Gaz (1877) 9:233

Faber, John Eberhard (1822-79). Lead pencil manufacturer,
    first to attach rubber tips and metallic point pro-
    tectors.

    Eng News (1879) 6:73

    DAB 6:241-2
    NYT (3/4/1879) 8:2

Fairbanks, Franklin (1828?-95). Mechanic and inventor
    of modifications of machinery used in scale making.

    Am Mach (1895) 18:368

    Appleton's Cyc 2:400

Fairbanks, Thaddeus (1796-1886). Inventor of machinery, refrigerator, and steam heating apparatus.

Am Manuf (5/7/1886) 38:9
RR Gaz (1886) 18:268

Appleton's Cyc 2:400
DAB 6:251

Fairlie, Robert Francis (? -1885). Inventor of Fairlie engine and narrow gauge system.

RR Gaz (1885) 17:536

Falconnet, Eugene F. (1833-87). Railroad engineer in South and West; inventions in aerial navigation.

ASCE Proc (1889) 15:135-6

Farmer, Moses Gerrish (1820-93). Inventor and electrician; pioneer in application of electricity to industry.

ASME Trans (1893) 14:1445-7
Elec World (1893) 21:402*
Eng & Min J (1893) 55:492
Eng Rec (1892-3) 27:506
Iron Age (1893) 51:1236
RR Gaz (1893) 25:416
Sci Am (1893) 68:338

DAB 6:279-81

Farnam, Henry (1803-83). Engineer and surveyor, several railroads and canals; designed and built first railroad bridge crossing the Mississippi.

ASCE Proc (1896) 22:618-22
ASCE Trans (1896) 36:605-8
RR Gaz (1883) 15:679

Appleton's Cyc 2:410
DAB 6:281
NYT (10/5/1883) 4:7

Farnham, John M. (? -1895). Mechanical engineer and inventor.

Eng Rec (1895) 32:399

Farnum, Henry Harrison (1808?-79). Engineer, canals.

RR Gaz (1879) 11:572

Faron, Edward (1824-89). Ship designer and engineer.

Eng (1889) 17:65*

Farquhar, Francis Ulric (1838-83). Military and civil engineer.

ASCE Proc (1889) 15:165
Assoc Eng Soc J (1882-3) 2:251-5
Eng News (1883) 10:320

NYT (7/6/1883) 2:1

Farra, Daniel (? -1897). Civil engineer and surveyor.

Eng Rec (1896-7) 35:135

Farrington, E. F. (1822?-98). Civil engineer, responsible for suspension bridge over Ohio River.

RR Gaz (1898) 30:316

Fava, Francis Renatus (? -1896). Civil engineer.

Eng Rec (1895-6) 33:309

Fay, John D. (1815?-95). Engineer and surveyor of bridges and canals.

Eng & Min J (1895) 59:563
Eng Rec (1895) 32:39
RR Gaz (1895) 27:395

Feeney, George W. (? -1896). Locomotive engineer.

RR Gaz (1896) 28:82

Feind, Bernhard (1849-94). Civil engineer, canals and water supply, Chicago.

ASCE Proc (1894) 20:182-3
Eng Rec (1894) 30:214
RR Gaz (1894) 26:603

Felton, S. K. (? -1897). Civil engineer.

Eng Rec (1897) 36:311

Felton, Samuel Morse (1809-89). Civil and railroad
engineer.

Am Eng & RR J (1889) 63:145
AISA Bull (1889) 23:28*
Am Manuf (2/15/1889) 44:11
ASCE Proc (1893) 19:92-4
Assoc Eng Soc J (1892) 11:199-215*
Eng Rec (1888-9) 19:126
Ind World (2/7/1889) 32:5
Iron Age (1889) 43:205-6
Nat Car (1889) 20:30
RR Gaz (1889) 21:78

Appleton's Cyc 2:428
DAB 6:318-9

Ferris, George Washington Gale (1859-96). Civil engineer,
designer of Ferris wheel.

Am Mach (1896) 19:1127
Am Manuf (1896) 59:766
Eng & Min J (1896) 62:515
Eng Rec (1896) 34:479
Ind World (11/26/1896) 47:5
Iron Age (1896) 58:1019
RR Gaz (1896) 28:835

DAB 6:339-40

Fessenden, John W. (1801-83). Civil and military engineer, railroads.

RR Gaz (1883) 15:115

NYT (2/10/1883) 2:3

Festestics, Charles Albert (? -1891). Civil engineer,
railroads.

Am Eng & RR J (1891) 65:381

Field, Burr Kellogg (1856-98). Civil engineer, railroads
and bridges; manufacturer of iron bridges.

Am Mach (1898) 21:77
ASCE Proc (1898) 24:777-9
ASCE Trans (1898) 40:568-70
ASME Trans (1898) 19:975-7
Elec Eng (1898) 25:81
Elec World (1898) 31:738
Eng Rec (1897-8) 37:139,161
Iron Age (1/20/1898) 61:21
RR Gaz (1898) 30:49

Field, Cyrus West (1819-92).   Promoter of first Atlantic
cable.

Am Mach (7/21/1892) 15:8
Elec Eng (1892) 14:64
Elec World (1892) 20:52*
Eng & Min J (1892) 54:60
Eng Rec (1892) 26:106
Pop Sci (1892-3) 42:287
Sci Am (1892) 67:53*

Appleton's Cyc 2:448-9*
DAB 6:357-9
NYT (7/13/1892) 5:1

Field, John (? -1898). Civil engineer of N.Y. and Ohio.

Eng Rec (1898) 38:553

Fink, Albert (1827-97). Civil engineer, inventor of
bridge truss; originator of trunk line pooling system.

ASCE Proc (1898) 24:859-72
ASCE Trans (1899) 41:626-38
Eng & Min J (1897) 63:359
Eng News (1897) 37:215*
Eng Rec (1896-7) 35:399
RR Gaz (1897) 29:262-3*

Appleton's Cyc 2:459*
DAB 6:387-8
NYT (4/4/1897) 5:4

Firth, Abraham (1818?-86). Superintendent of Boston &
Worcester Railroad; developer of improved methods of
railroad signalling.

RR Gaz (1886) 18:515

Firth, F. R. (1847?-72). Civil engineer, railroads.

RR Gaz (1872) 4:293

Fisher, Charles Henry (1835?-88). Civil engineer, chief
engineer of New York Central & Hudson River Railroad.

Am Eng & RR J (1888) 62:94
ASCE Proc (1893) 19:66-7
Eng & Min J (1888) 45:57
RR Gaz (1888) 20:61
Sanit Eng (1887-8) 17:124

Fisher, Henry Augustus (? -1898). Civil engineer.

Eng Rec (1897-8) 37:381

Fisher, Hiram H. (1832-89). Civil engineer and manufacturer of iron pipes in Allentown, Pa.

Iron Age (1889) 44:886

Fisher, P. D. (? -1889). Railroad engineer.

RR Gaz (1889) 21:86

Fisher, Robert Andrews (1832-93). Industrial chemist;
developer of sulphite process of paper making.

Frank Inst J (1894) 137:56

Fiske, Alfred R. (1820?-86). Consulting engineer of
New York, New England, and Western Investment Company.

RR Gaz (1886) 18:857

Fitch, Henry (1817?-95). Civil engineer, railroads and
bridges.

Eng & Min J (1895) 59:371
Eng Rec (1894-5) 31:363

Fitch, William H. (1825-96). Manufacturer of bridges.

Eng Rec (1896-7) 35:113
Iron Age (1/14/1897) 59:19

Fitton, Walter H. (? -1896). Chief engineer of
Cascade River locks in Oregon.

Eng Rec (1895-6) 33:165

Fitts, James H. (1861-93). Mechanical engineer.

ASME Trans (1893) 14:1449
Eng Rec (1893) 28:182

Flach, Emil J. (1864-93). Draftsman.

ASME Trans (1894) 15:1193-4

Flad, Henry (1824-98). Civil engineer, several railroads;
inventor of water and filter meters, electro-magnetic
air brakes, etc.

Am Eng & RR J (1898) 72:270
ASCE Proc (1899) 25:949-54
ASCE Trans (1899) 42:561-6
Assoc Eng Soc J (1899) 22:1-11
Eng & Min J (1898) 66:14
Eng News (1886) 16:9-10*
Eng Rec (1898) 38:69
RR Gaz (1898) 30:466,477*

DAB 6:445

Fleming, Francis H. (? -1894). Engineer of Central Park
and Hoosac Tunnel.

Eng Rec (1894) 30:86

Fleming, L. J. (? -1880). Civil engineer.

RR Gaz (1880) 12:248

Flint, Edward Austin (1832-86). Civil engineer, railroads in West and Mexico.

ASCE Proc (1886) 12:114-6

Flower, Joseph E. (1820?-99). Machinist; installed first
rotary presses in printing offices of N.Y. City.

Am Mach (1899) 22:923

Flynn, John H. (1828?-84). Master railroad mechanic of
Western & Atlantic Railroad.

Am Mach (10/18/1884) 7:8
Nat Car (1884) 15:144

Flynn, Patrick John (1838-93). Civil engineer, bridges
and public works in India, irrigation works in Calif.

Am Eng & RR J (1893) 67:355
ASCE Proc (1894) 20:68-9
Eng & Min J (1893) 55:588
RR Gaz (1893) 25:442

Fobes, Richard (1858-91). Civil engineer of Worcester,
Mass.

Assoc Eng Soc J (1894) 13:136-7

Focht, George (1823?-98). Inventor of self-dumping coal
tub; iron manufacturer.

Am Mach (1898) 21:491-2
Eng Rec (1898) 38:69
Iron Age (6/23/1898) 61:21

Fogg, Charles E. (1825-91). Civil engineer, water works
and railroads in N.Y. State and New England.

ASCE Proc (1891) 17:230-1

Foote, Emerson (? -1858). Railroad superintendent of
Georgia Central Railroad.

Am Eng & RR J (1858) 31:653

Ford, Arthur Livermore (1851-80). Civil engineer and mineralogist; railroads in Latin America.

ASCE Proc (1880) 6:75-6

Ford, George E. (? -1893). Civil engineer, railroads.

Eng Rec (1892-3) 27:453

Ford, James K. (1816?-99). Civil engineer, railways.

Eng Rec (1899) 40:37

Forshey, Caleb Goldsmith (1812-81). Civil engineer, instructor of civil engineering and meteorology; responsible for improvements on Mississippi.

ASCE Proc (1897) 23:173-4
ASCE Trans (1897) 37:560-1

Appleton's Cyc 2:506

Forster, John Harris (1822-94). Civil engineer, surveyor of boundaries for Great Lakes and Mexico.

Eng & Min J (1894) 57:611
Eng Rec (1894) 30:52

Foster, Theodore Reno (1854-97). Mechanical engineer, master mechanic of Denver & Rio Grande Railroad.

ASME Trans (1897) 18:1104

Fowle, Andrew S. (1821?-99). Maker of first engraving machine for U.S. notes.

Am Mach (1899) 22:222

Fowler, Charles Edward (1841-83). Engineer and draftsman; city engineer of New Haven, Conn.

ASCE Proc (1889) 15:164

Fowler, Hiram (? -1891). Civil engineer of New England.

Am Eng & RR J (1891) 65:139
RR Gaz (1891) 23:103

Fowler, Lucien D. (? -1898). Engineer of street and water board in Jersey City, N.J.

Eng Rec (1897-8) 37:139

Fowler, Thaddeus (? -1886). Inventor and manufacturer.

Am Manuf (12/31/1886) 39:6

Fox, Jesse W. (1819?-94). Engineer and surveyor for railroads in West.

Eng & Min J (1894) 57:348

Fox, Joseph G. (1843-90). Professor of civil and topographical engineering at Lafayette College.

Eng & Min J (1890) 49:32

Francis, James (1840-98). Hydraulic engineer, surveyor for Hoosac Tunnel.

ASME Trans (1899) 20:1003
Assoc Eng Soc J (1899) 22:12-4
Eng Rec (1898-9) 39:39
RR Gaz (1898) 30:884

Francis, James Bicheno (1815-92). Hydraulic engineer, locomotive builder; designer of turbine of mixed flow.

Am Eng & RR J (1892) 66:479
ASCE Proc (1893) 19:74-7*
Assoc Eng Soc J (1894) 13:1-9*
Eng & Min J (1892) 54:301
Eng News (1887) 17:14; (1892) 28:266*; (1894) 32:28-30
Eng Rec (1892) 26:258
Iron Age (1892) 50:532
RR Gaz (1892) 24:714

Appleton's Cyc 2:522-3*
DAB 6:578-9

Francis, Joseph (1801-93). Inventor and builder of life boats; pioneer in use of corrugated metal in their construction.

Sci Am (1893) 68:310

Appleton's Cyc 2:524*
DAB 6:582-3

Frederick, Thomas W. (1857?-86). Master mechanic.

RR Gaz (1886) 18:730

Free, Benjamin (? -1891). Locomotive engineer.

Nat Car (1891) 22:42

Freeman, Ernest G. (? -1900). Bridge engineer.

Eng Rec (1900) 41:233
RR Gaz (1900) 32:177

Freeman, H. C. (1829-1900). Mining engineer.

Eng & Min J (1900) 70:586

Frehsee, Julius (? -1896). Civil engineer and surveyor of Western N.Y. State.

Eng Rec (1896) 34:457

Fremaux, L. J. (? -1898). Civil engineer and architect.

Eng Rec (1897-8) 37:535

Fremont, T. C. (1816?-86). Civil engineer, railroads.

RR Gaz (1886) 18:340

French, Edmund (1807?-60). Civil engineer, assistant engineer of Croton Aqueduct.

ASCE Proc (1897) 23:170-1
ASCE Trans (1897) 37:561-2

French, Edward Clement (1858-95). Superintendent and manager of Deseronto Chemical Works.

ASME Trans (1895) 16:1197

Frick, George (1826-92). Inventor and manufacturer of Frick engine.

Am Eng & RR J (1893) 67:101-2
Am Mach (1/12/1893) 16:8
Eng Rec (1892-3) 27:92

Frick, Joseph K. (? -1886). Engineer and architect.

Sanit Eng (1885-6) 13:352

Frisbie, D. (? -1889). Manufacturer of passenger and freight elevators.

Sci Am (1889) 61:144

Frisbie, Eaton N. (1830?-93). Civil engineer, surveyor for Lehigh Valley Railroad.

Eng & Min J (1893) 56:550

Frisbie, Russel (1822?-98). Pattern maker; president of J. & E. Stevens Company.

Iron Age (2/3/1898) 61:27

Frishmuth, William (1823?-93). Inventor of process for making aluminum and nickel plating.

AISA Bull (1893) 27:234

Fritz, George (? -1873). Mechanical engineer of Cambria
    Iron Works.

    AISA Bull (1873) 7:396
    Van Nostrand's (1873) 9:445-7

Fritz, William H. (? -1884). Machinist and manufacturer
    of pig iron.

    Eng & Min J (1884) 37:273

Frobell, B. W. (1830-88). Military engineer.

    Am Eng & RR J (1888) 62:428
    Eng Rec (1888) 18:118
    RR Gaz (1888) 20:481

Froelich, John C. (1848?-98). Manufacturer of machinery.

    Iron Age (4/14/1898) 61:17

Frost, Benjamin Dix (1830-80). Civil engineer, chief
    engineer for Hoosac Tunnel.

    ASCE Proc (1887) 13:139-40
    Eng News (1880) 7:248

Fry, Edward (? -1890). Chief mechanical engineer of
    Brooklyn.

    Eng Rec (1890) 22:66

Fry, Howard (1847-83). Master mechanic, superintendent of
    motive power on several railroads; locomotive designer.

    Am Mach (5/26/1883) 6:2
    RR Gaz (1883) 15:291-2*

Fuller, Aspinwall (? -1888). Marine engineer.

    Eng (1888) 16:7

Fuller, Charles (? -1890). Military engineer.

    Eng Rec (1890-1) 23:38

Fuller, Charles E. (? -1894). Builder of first wooden
    bridge across Mississippi.

    Eng Rec (1893-4) 29:328

Fuller, Levi Knight (1841-96). Mechanical engineer and manufacturer of wood working machinery; inventor of appliances for organs.

ASME Trans (1897) 18:1093-4
Iron Age (1896) 58:779

DAB 7:59

Fulton, Robert (1765-1815). Civil engineer and inventor of steamships.

Mech Mag (1834) 4:366-76

DAB 7:68-72

Furlong, Francis F. (? -1896). Mining engineer.

Eng Rec (1896) 34:457

Galloway, William (1809-90). Locomotive engineer.

Eng & Min J (1890) 49:428
Iron Age (1890) 45:605
Nat Car (1890) 21:74

Gallup, Benjamin Ela (? -1895). Civil engineer, railroads.

Eng Rec (1895-6) 33:2

Gamewell, John N. (? -1896). Inventor of Gamewell fire alarm.

Elec Eng (1896) 22:114

Gardanier, George W. (? -1900). Chief electrical engineer of Western Union.

Elec World (1900) 36:708

Gardner, Arthur B. (? -1895). Civil engineer.

Eng Rec (1894-5) 31:273

Gardner, Franklin (1820?-98). Manufacturer of steam engines and cars.

Am Mach (1898) 21:909

Gardner, H. A. (1816-75). Civil engineer, constructor of several railroads in Midwest.

ASCE Proc (1873-5) 1:335-7
Eng News (1875) 2:106
RR Gaz (1875) 7:318

NYT (7/28/1875) 4:6

Gardner, William (1835-98). Gas engineer.

AISA Bull (1898) 32:123
Eng Rec (1898) 38:223

Garfield, Edwin (1816?-77). Locomotive engineer.

RR Gaz (1877) 9:522

Garnett, Charles F. M. (1810?-86). Railroad engineer.

RR Gaz (1886) 18:220

Garrison, Daniel R. (1815?-96). Pioneer in manufacture of steam engines and heavy machinery in St. Louis.

Eng & Min J (1896) 61:259
Eng Rec (1895-6) 33:255
RR Gaz (1896) 28:207

Garsed, Richard (1819?-97). Manufacturer and mechanic of Philadelphia.

Am Mach (1897) 20:593

Garvin, H. R. (1831?-87). Manufacturer of machinery and machinist's tools.

Am Mach (1/14/1888) 11:7

Gaskill, Harvey Freeman (1845-89). Mechanical engineer and inventor of Gaskill pumping engine.

Am Eng & RR J (1889) 63:243
Am Mach (4/11/1889) 12:8-9
ASME Trans (1889) 10:833
Eng & Min J (1889) 47:329
Eng Rec (1888-9) 19:250,278
Iron Age (1889) 43:555
Sci Am (1889) 60:245

DAB 7:177-8

Gass, J. B. (? -1899). City and sanitary engineer of Fort Smith, Arkansas.

Eng Rec (1899) 40:734

Gatchell, George B. (? -1889). Engineer and superintendent of railroads.

RR Gaz (1889) 21:561

Gates, Justin J. (? -1896). Manufacturer of incandescent lamps.

Elec Eng (1896) 21:392

Gaty, Samuel (? -1887). Machinist; first engineer west
of Mississippi.

Am Eng & RR J (1888) 61:298

Gause, J. Taylor (1823-99). Shipbuilder.

Am Eng & RR J (1899) 73:30

Geary, Michael (1844?-95). Manufacturer of engines;
president of Oil City Tube Company and Oil City Boiler
Works.

Am Manuf (1895) 57:447
Iron Age (1895) 56:644

Geddes, George (1809-83). Civil engineer, canals and
railroads.

RR Gaz (1883) 15:679

Appleton's Cyc 2:621
NYT (10/9/1883) 2:2

Geer, Herbert Guernsey (1869-1900). Mechanical and
consulting engineer; professor of mechanical engineering.

ASME Trans (1900) 21:1165-6
Elec World (1900) 35:420
Eng Rec (1900) 41:281

Geyelin, Emile C. (1825-1900). Engineer and designer of
hydraulic power plants.

Am Mach (1900) 23:628
Elec World (1900) 36:42
Eng Rec (1900) 41:625
Iron Age (6/28/1900) 65:20

Gibboney, John W. (? -1897). Inventor of electrical
applications.

Eng Rec (1897-8) 37:91

Gibbs, Martin L. (1837?-96). Inventor of Gibbs & Champion plow.

Iron Age (1896) 58:970

Giblin, Arthur Leon (1866-94). Civil engineer, instructor of civil engineering.

ASCE Proc (1894) 20:171-2

Gilbert, John S. (? -1891). Naval architect and inventor.

Eng Rec (1891) 24:184

Gilbert, Rufus Henry (1832-85). Inventor; projector of elevated railway system in N.Y.

Iron Age (7/16/1885) 36:17
RR Gaz (1885) 17:461
Sci Am (1885) 53:69

Appleton's Cyc 2:646-7
DAB 7:271-2

Gilbert, William Bradford (1810?-97). Civil engineer, several early railroads.

RR Gaz (1897) 29:696

Gill, Joseph (? -1897). Civil engineer of West Virginia.

Eng Rec (1896-7) 35:531

Gillam, Sewal (? -1896). Civil engineer and ironmaster.

Eng Rec (1896) 34:271

Gillespie, Joshua Lathrop (1848-90). Civil engineer; preservation of Falls of St. Anthony.

ASCE Proc (1890) 16:224-5

Gillespie, William Mitchell (1816-68). Civil engineer; first professor of civil engineering at Union College.

Sci Am (1868) 18:42

Appleton's Cyc 2:651
DAB 7:288-9

Gillet, James (1818-88). Editor of National Car and Locomotive Builder.

Am Mach (6/2/1888) 11:8
Nat Car (1888) 19:94-5*
RR Gaz (1888) 20:325

Gillham, Robert (1854-99). Civil engineer; instituted cable traction for street railways.

ASCE Proc (1900) 26:101-5
Eng Rec (1899) 40:18
RR Gaz (1899) 31:378

Gilliss, John Roberts (1842-70). Civil engineer, Central Pacific and Union Pacific Railroads.

ASCE Proc (1896) 22:690-1
ASCE Trans (1896) 36:555-6

Gillmore, Quincy Adams (1825-88). Civil and military engineer; author of several books and treatises on engineering.

ASCE Proc (1894) 20:60-1
Sci Am (1888) 58:273
Van Nostrand's (1872) 7:1-10*

Appleton's Cyc 2:653-4*
DAB 7:295
NYT (4/8/1888) 1:4

Gilmore, Robert J. (1847-96). Mechanical engineer, machine and brass foundry business.

ASME Trans (1897) 18:1090
Eng Rec (1896) 34:101

Githens, Joseph C. (1835?-1900). Inventor of Little Giant rock drill.

Am Mach (1900) 23:45
Eng Rec (1900) 41:42

Gleason, John (1826?-98). Manufacturer and inventor of lathe for turning gun stocks.

Am Mach (1898) 21:209
Iron Age (3/17/1898) 61:27

Glenn, John S. (1845?-95). Inventor of Glenn air brake, valves, etc.

Iron Age (1895) 56:697

Glenn, John W. (1806?-92). Military engineer in Confederate Army.

Eng Rec (1891-2) 25:326

Goebel, Henry (1818-93). Developer of electrical devices and incandescent lamps.

Elec World (1893) 21:333*; (1893) 22:456*
Eng Rec (1893-4) 29:34
Sci Am (1894) 70:6

Goetz, George Washington (1855-97). Metallurgical engineer.

AIME Trans (1897) 27:436-44
Eng & Min J (1897) 63:203*
Eng Rec (1896-7) 35:157

DAB 7:358

Golay, Philip (? -1898). Engineer in charge of locks and dams.

Eng Rec (1898-9) 39:3

Goldsborough, John McDowell (? -1895). Civil engineer, railroads.

Eng Rec (1895-6) 33:75

Goodell, Austin W. (1842?-1900). Inventor and manufacturer of woodworking machinery.

Am Mach (1900) 23:360
Iron Age (4/19/1900) 65:32

Gooding, William (1803-78). Chief engineer of Illinois & Michigan Canal.

Eng News (1878) 5:83

Goodwillie, James Barnett (1873-98). Mining and metallurgical engineer.

AIME Trans (1899) 29:xxix-xxx

Goodwin, Homer Stanley (1832-92). Civil engineer, several railroads; instructor at Lehigh University.

Am Eng & RR J (1893) 67:101
ASCE Proc (1893) 19:163-5
Eng & Min J (1892) 54:636
Eng Rec (1892-3) 27:92
Nat Car (1893) 24:10
RR Gaz (1892) 24:997

Goodwin, John Marston (1833-91).   Civil engineer; inventor of coal cars and boats.

Am Eng & RR J (1891) 65:574
Am Manuf (1891) 49:819
ASCE Proc (1891) 17:267-70
Eng Rec (1891) 24:360
RR Gaz (1891) 23:790

Goodwin, William A. (? -1896).   Civil engineer, railroads; city engineer of Portland, Me.

Eng Rec (1895-6) 33:309

Goodyear, Charles (1800-60).   Discoverer of the process
of vulcanization of rubber.

Pop Sci (1898) 53:690-700*
Sci Am (1869) 21:408

Appleton's Cyc 2:683-4*
DAB 7:413-5

Goodyear, Miles W. (? -1889).   Electrician.

Elec Eng (1890) 9:33
Elec World (1889) 14:410

Gordon, Charles (? -1899).   Designer and manufacturer
of heating apparatus.

Eng Rec (1899) 40:658

Gordon, Edward (1813?-77).   Locomotive engineer.

RR Gaz (1877) 9:479

Gorlinski, Joseph (1825?-1900).   Government surveyor and
engineer.

Eng Rec (1900) 42:234

Gorringe, Henry Honeychurch (1840-85).   Shipbuilder and
consulting engineer; responsible for moving
Cleopatra's needle to N.Y.

ASCE Proc (1890) 16:215-6
ASME Trans (1885) 6:875
Iron Age (7/9/1885) 36:33

Appleton's Cyc 2:689-90
DAB 7:437
NYT (7/7/1885) 5:1

Gottlieb, Abraham M. (1837-94). Bridge engineer of Chicago; chief engineer of World Columbian Exposition.

ASCE Proc (1894) 20:76-8
Assoc Eng Soc J (1894) 13:233-5
Eng Rec (1893-4) 29:184
Iron Age (1894) 53:320
RR Gaz (1894) 26:130

Gould, James H. (1844-97). Manufacturer of pumps and hydraulic machinery.

Elec World (1897) 29:80
Ind World (1/14/1897) 48:5
Iron Age (1/14/1897) 59:20

Gowen, John E. (1825-95). Marine engineer.

Eng (1895) 29:131

Graff, Christopher L. (1823?-97). Iron manufacturer.

AISA Bull (1897) 31:45
Am Manuf (1897) 60:263
Eng & Min J (1897) 63:215
Iron Age (2/25/1897) 59:20

Graff, Frederic (1817-90). Civil engineer, expert on establishment of water works.

Am Eng & RR J (1890) 64:238
ASCE Proc (1891) 17:247-50
Eng Rec (1889-90) 21:274
Frank Inst J (1890) 129:517-21
Iron Age (1890) 45:558

DAB 7:466-7

Graham, Augustus Clason (1828-92). Co-founder of the Electrician (later Electrical Engineer).

Elec Eng (1892) 14:262

Graham, Charles Kinnaird (1824-89). Military and civil engineer.

Am Eng & RR J (1888) 62:244
Eng Rec (1888-9) 19:278

Appleton's Cyc 2:700-1
DAB 7:471

Graham, John (? -1894). Engineer for Central Park and sewers in Columbus.

Eng Rec (1893-4) 29:248

Grant, David Beach (? -1888). Manufacturer of locomotives.

Am Eng & RR J (1888) 62:333

Grant, John A. (1866?-97). Mining engineer.

Eng & Min J (1897) 63:71

Grant, McGee (1822-92). Railroad engineer in Northeast.

Assoc Eng Soc J (1893) 12:544-5

Grant, William Harrison (1815-96). Civil engineer and
landscape architect, surveyor for several railroads.

ASCE Proc (1896) 22:610-1
ASCE Trans (1896) 36:557-8
Eng & Min J (1896) 62:371
Eng Rec (1896) 34:361
RR Gaz (1896) 28:727

Grant, Winfield H. E. (1861-92). Consulting engineer,
specialist in ventilation and heating and drying
apparatus.

ASME Trans (1892) 13:675-6

Gray, Hugh (1807-85). Master car builder.

RR Gaz (1885) 17:604

Gray, Joshua (1824-99). Inventor of quick steaming boiler
tube, breech loading repeating rifle, railroad signal,
etc.

Am Mach (1899) 22:624
Eng Rec (1899) 40:110
Iron Age (6/29/1899) 63:21

Green, George F. (? -1892). Inventor in the line of
electric railways.

Elec World (1892) 19:21-2*,410

Green, Norvin (1818-93). One of the founders of AIEE;
president of Western Union.

AIEE Trans (1893) 10:665-6
Elec Eng (1893) 15:174
Elec World (1893) 21:118*
RR Gaz (1893) 25:136

Appleton's Cyc 2:745
DAB 7:555
NYT (2/13/1893) 1:7

Green, William H, (? -1893). Steam engineer and manufacturer.

Iron Age (1893) 51:1070

Greene, A. S. (? -1896). Civil and military engineer of U.S. Navy.

Am Mach (1896) 19:331
Eng & Min J (1896) 61:259
Eng Rec (1895-6) 33:273

Greene, Benjamin Henry (1830-90). Railroad engineer of South and West.

ASCE Proc (1890) 16:187-8
Assoc Eng Soc J (1891) 10:105-13

Greene, George Sears (1801-99). Civil engineer, several railroad and water works projects; one of the founders of ASCE.

Eng & Min J (1899) 67:150
Eng News (1899) 41:71*
Eng Rec (1898-9) 39:202

Appleton's Cyc 2:749
DAB 7:566

Greene, Samuel Dana (1839?-1900). Electrical engineer; associate of Edison.

AIEE Trans (1899) 16:691-5
Am Manuf (1900) 66:47
Elec World (1900) 35:52*
Iron Age (1/11/1900) 65:29

Appleton's Cyc 2:749

Greenland, Walter W. (? -1895). Civil engineer.

Eng Rec (1894-5) 31:309

Greenwood, Miles (1807-85). Manufacturer of hydraulic presses, steam engines, iron fronts for buildings, and heating appliances.

RR Gaz (1885) 17:733

Appleton's Cyc 2:758
DAB 7:592-3

Greenwood, William H. (1832-80). Civil and topographical engineer, railroads near Mexico.

ASCE Proc (1881) 7:89-90
Eng News (1880) 7:303

Gregory, Henry Payson (1841-88). Member of Corps of Engineers; operator of machinery business.

ASME Trans (1889) 10:834

Gregory, James F. (? -1897). Member of Corps of Engineers.

Eng Rec (1897) 36:201

Grieg, James (? -1897). Civil engineer.

Eng Rec (1896-7) 35:531

Gridley, Vernon Hill (1867-96). Civil engineer, public works, including street pavements in Brooklyn.

ASCE Proc (1896) 22:577
ASCE Trans (1896) 36:595-6
Eng Rec (1896) 34:289

Griffen, John (1812-84). Civil and mechanical engineer; improved manufacture of wrought iron beams.

AISA Bull (1884) 18:20
ASCE Proc (1886) 12:38
Eng & Min J (1884) 37:40
Iron Age (1/24/1884) 33:15
RR Gaz (1884) 16:55

Griffith, Benjamin P. (? -1899). Civil engineer, railroads.

RR Gaz (1899) 31:107

Griggs, George Henry (1832?-91). Master mechanic of several railroads.

Nat Car (1892) 23:10

Griggs, Gregory S. (1805?-70). Superintendent of motive power for Boston & Providence Railroad.

Nat Car (8/1870-1) 1:5

Griscom, William Woodnutt (1851-97). Electrical engineer.

Elec Eng (1897) 24:311
Elec World (1897) 30:402
Eng & Min J (1897) 64:462

Grist, James Edmund (1864-97). Machinist; superintendent of Pennsylvania Iron Works.

ASME Trans (1897) 18:1106
Iron Age (5/27/1897) 59:21

Griswold, John Augustus (1818-72). Manufacturer of
machinery, plates, and iron work for the Monitor.

Sci Am (1872) 27:312

Appleton's Cyc 3:3*
DAB 8:8-9
NYT (11/1/1872) 5:2

Grosscup, Manias G. (1838?-97). Inventor of improved
farming machinery.

Iron Age (6/24/1897) 59:16

Guild, William H. (1832-85). Mechanical engineer; manufacturer of steam pumps.

Am Mach (12/5/1885) 8:9
Sanit Eng (1885) 12:519
Sci Am (1885) 53:385

Guilford, Simeon (1801?-95). Civil engineer, canals and
furnaces.

Iron Age (1895) 55:391

Gurley, Lewis E. (1826?-97). Manufacturer of surveyor's
and engineer's instruments.

Eng & Min J (1897) 63:516

Gurley, William (? -1887). Manufacturer of mathematical
instruments; president of R.P.I.

Sanit Eng (1886-7) 15:166

Gwin, Walter (1802?-82). Military and civil engineer,
canals and public works.

RR Gaz (1882) 14:90

Haagensen, Sophus (? -1891). Civil engineer, U.S. Coastal survey.

Eng Rec (1891-2) 25:54

Haessler, Francis Joy (? -1900). Inventor of mechanical
and electrical devices for naval purposes.

Am Mach (1900) 23:1147
Eng Rec (1900) 42:500

Hailes, William (? -1892). Inventor of base-burning
stoves.

Eng & Min J (1892) 54:228

Hain, Franklin Kintzle (1846?-96). Mechanical engineer, railroads and locomotives.

Am Eng & RR J (1896) 70:121-2
Eng Rec (1895-6) 33:417
Iron Age (1896) 57:113
RR Gaz (1896) 28:347*

Hainsworth, William (1833-96). Inventor of hydraulic forging press and soaking pit.

AISA Bull (1896) 30:244
Iron Age (1896) 58:826

Hale, Albert W. (1839?-99). Civil engineer.

Eng & Min J (1899) 68:254

Hall, Ferdinand (1855-99). Civil engineer, railroads.

RR Gaz (1899) 31:834
West Soc Eng J (1899) 4:499-501*

Hall, George Thomas (1845-91). Civil engineer, railroads.

ASCE Proc (1891) 7:97-9

Hall, Thomas Seavey (1827-80). Inventor of automatic electric railway signals.

Sci Am (1880) 43:405

DAB 8:145-6

Hall, William C. (1858-97?). Engineer and constructor of water works in Boston.

Assoc Eng Soc J (1898) 20:63-5

Hall, Willis H. (1860-95). Draftsman.

Assoc Eng Soc J (1895) 15:40

Hallidie, Andrew Smith (1836-1900). Civil engineer, responsible for the first cable road in San Francisco.

Am Mach (1900) 23:432
Eng & Min J (1900) 69:524*
Eng Rec (1900) 41:430

DAB 8:156

Hallock, John Keese (1844-97). Inventor and patent lawyer.

ASME Trans (1897) 18:1103-4

Halm, W. G. (? -1891). Electrician.

Elec World (1891) 18:459

Halpin, William G. (? -1892). City engineer of Cincinnati.

Eng Rec (1891-2) 25:410

Hamblet, James (1824?-1900). Manufacturer of electrical apparatus and telegraph instruments.

Elec World (1900) 35:56*
Sci Am (1900) 82:59

Hamer, Thomas J. (1831?-85). Mechanical engineer.

RR Gaz (1885) 17:189

Hamilton, Alexander (1854-89?). Mechanical engineer.

ASME Trans (1889) 10:835

Hamilton, George W. (1816?-1900). Mechanic.

Am Mach (1900) 23:1032

Hamilton, James (? -1893). Railroad engineer of Chicago.

Eng Rec (1893-4) 29:2
RR Gaz (1893) 25:878

Hamilton, John Mark (? -1893). Civil engineer, professor in De La Salle Institute, N.Y. City.

Eng Rec (1892-3) 27:393

Hamlin, Frederick Hinman (1848-84). Engineer of public works in N.Y. City.

Sanit Eng (1884) 10:576

Hammer, Hakron (? -1896). Civil engineer.

Eng Rec (1896) 34:457

Hammond, Andrew R. (? -1896). Mining engineer in South Africa.

Eng Rec (1895-6) 33:380

Hampson, John (? -1854). Railroad engineer.

Am Eng & RR J (1854) 27:411

Handren, John W. (1832-92). Marine engineer, manufacturer of marine engines.

Am Eng & RR J (1892) 66:479
ASME Trans (1892) 13:686
Eng (1892) 24:66

Hanscom, William Wallace (1839-88). Consulting engineer;
constructor of cable railways, electric lighting, and
water works.

Am Eng & RR J (1888) 62:190
ASME Trans (1888) 9:737

Hardee, Thomas P. (? -1880). Civil engineer; chief
engineer of Louisiana.

Iron Age (5/27/1880) 25:26

Hardick, Charles B. (? -1874). Mechanical engineer;
inventor of Niagara direct acting steam pump.

Eng & Min J (1874) 18:260

Harding, Horace (1827?-99). Civil engineer, locks and
dams.

ASCE Proc (1899) 25:1059-60
Eng & Min J (1899) 68:164

Hare, Adam (1829-98). River engineer and inventor.

Am Mach (1898) 21:848

Harlan, Samuel (1806-83). Builder of first American iron
steamship.

Iron Age (2/8/1883) 31:15

NYT (2/7/1883) 5:5

Harper, Deratus (? -1878). Civil engineer; builder of
first float-bridge, swing bridge, and iron bridge in
Chicago.

Eng News (1878) 5:145

Harris, Benjamin Marvin (1866-96). Mechanical and
electrical engineer.

ASME Trans (1897) 18:1089

Harris, Clarendon (1836?-92). Civil engineer, railroads.

Eng Rec (1891-2) 25:394

Harris, Daniel Lester (1818-79). Civil engineer, several railroads.

RR Gaz (1879) 11:389,404

DAB 8:307
NYT (7/13/1879) 7:2

Harris, Elisha (1807-90). Inventor of some improvements in cotton mill.

Iron Age (1890) 46:257

Harris, Henrique (1837-82). Civil engineer, public works; chief engineer of New York & Manhattan Beach Railway.

ASCE Proc (1897) 23:175
ASCE Trans (1897) 37:562
Eng News (1882) 9:367

Harris, Joel Benedict (1822-91). Civil engineer, manufacturer of wheels and castings.

Am Eng & RR J (1891) 65:524
Iron Age (1891) 48:693
RR Gaz (1891) 23:772

Harris, John Albert (? -1898). Civil engineer with Chesapeake & Ohio Railway.

Eng Rec (1897-8) 37:557

Harris, John H. (1838-94). Responsible for direct action steam pump.

Am Mach (2/1/1894) 17:8
ASME Trans (1894) 15:1188-9
Cassier's (1894) 6:84-6
Eng (1894) 27:31
Eng Rec (1893-4) 29:134

Harris, Robert (1831?-94). Civil engineer, vice president of Northern Pacific Railroad Company.

Eng & Min J (1894) 57:396
Eng Rec (1893-4) 29:345

Harris, Robert Lewis (1834?-96). Civil engineer, railroads.

Eng & Min J (1896) 62:347
Eng Rec (1896) 34:325
Iron Age (1896) 58:683
RR Gaz (1896) 28:710

Harris, William Andrew (1835-96). Steam engineer, builder of Harris-Corliss steam engine.

Eng Rec (1896) 34:399
Iron Age (1896) 58:871
Power (9/1896) 16:23

Harrison, Joseph (1810-74). Mechanical and railroad engineer, constructor of steam boilers and locomotives.

Frank Inst J (1874) 97:377-80
Sci Am (1874) 30:248

Appleton's Cyc 3:100
DAB 8:345-6

Harrison, Joseph H. (1835-96). Manufacturing engineer.

Iron Age (1896) 57:979

Harrison, William (? -1889). Railroad engineer.

Am Eng & RR J (1889) 63:243

Harson, William (1812-85). Engineer and inventor of compressed air rock drill.

Iron Age (4/16/1885) 35:19

Hartley, Roger (1828-97). Railroad engineer and coal operator of Pittsburgh.

Am Manuf (1897) 60:624

Hartupee, Andrew (1819-91). Engineer and machinist; inventor of first compound engine and nut making engine.

Iron Age (1891) 48:500

Harvey, Hayward Augustus (1824-93). Inventor of screw machinery and Harveyized steel armor plate process.

Am Eng & RR J (1893) 67:501
Am Manuf (1893) 53:356
Iron Age (1893) 52:391
RR Gaz (1893) 25:661
Sci Am (1893) 69:178; (1897) 77:133

DAB 8:373-4

Hasbrouck, Robert M. (1823?-86). Civil engineer, Erie Canal.

RR Gaz (1886) 18:65

Haskell, Charles Frederick Beale (1856-95). Civil engineer, railroads in Midwest.

ASCE Proc (1895) 21:182
Eng Rec (1895) 32:3
RR Gaz (1895) 27:374

Haskell, James R. (1832?-97). Inventor of Lyman-Haskell multi-charge gun.

Am Mach (1897) 20:649-50
Eng Rec (1897) 36:245

Haskin, DeWitt Clinton (1824-1900). Railroad engineer and tunnel builder.

Eng Rec (1900) 42:90
Iron Age (8/9/1900) 66:21
RR Gaz (1900) 32:545
Sci Am (1900) 83:99

Haskins, John Ferguson (1833-93). Mechanical engineer.

ASME Trans (1893) 14:1444-5

Haslett, Sullivan (1844?-87). Civil engineer, railroads and bridges; consulting engineer for Chautauqua Lake Railroad.

ASCE Proc (1887) 17:127-8
RR Gaz (1887) 19:13
Sanit Eng (1886-7) 15:142

Hatch, Charles F. (? -1889). Railroad engineer of Midwest.

Am Eng & RR J (1889) 63:244

Hatfield, Robert G. (1815-79). Architectural engineer.

ASCE Proc (1896) 22:614-5
ASCE Trans (1896) 36:558
Sci Am (1879) 40:177

NYT (2/20/1879) 2:1

Hatton, James Edwin (? -1885). Locomotive engineer.

RR Gaz (1886) 18:12

Haughan, Charles P. (1842?-1900). Improver of process for production of chrome steel.

AISA Bull (1900) 34:92
Am Mach (1900) 23:436
RR Gaz (1900) 32:311

Hawley, A. W. (? -1886). Engineer and master mechanic of New York, New Haven & Hartford Railroad.

RR Gaz (1886) 18:205

Hawn, Frederick (1810-98). Civil engineer and surveyor of Kansas.

Eng & Min J (1898) 65:228
Eng Rec (1897-8) 37:249

Hay, A. T. (? -1895). Inventor and metallurgist; constructor of first all steel bridge across Missouri River.

Eng Rec (1894-5) 31:183

Hay, Alexander (1813?-84). Railroad engineer and canal builder.

Iron Age (5/15/1884) 33:34

Hay, S. F. (? -1895). Mechanical engineer.

Eng Rec (1894-5) 31:237

Hayden, Edward Simeon (? -1899). Metallurgist and inventor of Hayden process for separating precious metals from copper.

Eng & Min J (1899) 67:240
Iron Age (1899) 63:18

Hayden, Henry A. (? -1895). Railroad engineer.

Eng Rec (1895-6) 33:21

Hayes, Samuel J. (1816-82). Mechanic and inventor of improvements in railroads.

RR Gaz (1882) 14:591-2*

Hayward, James A. (1849-80). Member of U.S. Corps of Engineers; canals and harbor work.

ASCE Proc (1881) 7:88

Haywood, Benjamin (1792-1878). Manufacturer of mining machinery and steam engines.

AISA Bull (1878) 12:164
Iron Age (7/18/1878) 22:1

Appleton's Cyc 3:148

Hazlehurst, Henry (1815-1900). Civil and marine engineer.

Eng Rec (1900) 41:209
Iron Age (3/8/1900) 65:19

Heermans, Thomas (1829-92). Civil engineer, railroads in
   N.Y. State.

   Eng Rec (1892-3) 27:48
   Iron Age (1892) 50:1117

Hegeman, Allen B. (1861-92). Railroad engineer in East
   and Mexico.

   ASCE Proc (1893) 19:171

Heilman, Oren Gibson (1866-94). Instructor of mechanical
   engineering at Sibley College.

   Am Mach (8/2/1894) 17:8
   ASME Trans (1895) 16:1187-8
   Eng & Min J (1894) 58:59
   Eng Rec (1894) 30:118
   RR Gaz (1894) 26:526

Heinrich, Oswald J. (1827-86). Mining engineer and
   architect.

   AIME Trans (1885) 14:784-9
   Eng & Min J (1886) 41:105,133

Helme, William (1824-88). Manufacturer of gas meters
   and car wheels.

   Frank Inst J (1888) 126:155-7

Hemenway, Frank F. (1837-98). Inventor of Hemenway
   automatic cutoff engine.

   Am Eng & RR J (1898) 72:384
   Am Mach (1898) 21:789
   ASME Trans (1899) 20:1000
   Eng (11/1/1898) 35:4
   Eng Rec (1898) 38:443
   Mach (11/1898-9) 5:76
   Power (11/1898) 18:17

Hemphill, James J. (1827-1900). Mechanical engineer;
   designer and manufacturer of steam engines and heavy
   machinery.

   AISA Bull (1900) 34:139
   Am Mach (1900) 23:801
   Am Manuf (1900) 67:106*
   Elec World (1900) 36:268
   Eng & Min J (1900) 70:196
   Eng Rec (1900) 42:138
   Iron Age (1889) 44:212; (8/9/1900) 66:22
   RR Gaz (1900) 32:559

Hennessey, William A. (? -1897).   Inventor of Hennessey
    triple draft tubular boiler.

    Eng Rec (1896-7) 35:355
    Iron Age (3/25/1897) 59:18

Henning, Benjamin S. (? -1900).   Railway engineer, known
    for scheme of gravity tunnels between N.Y. City and
    Brooklyn.

    Eng Rec (1900) 42:138

Henry, Benjamin Tyler (1821?-98).   Inventor of Winchester
    repeating rifle.

    Iron Age (6/16/1898) 61:34

Henry, Frank (1821-96).   Inventor of foot hammer and other
    sewing machine attachments.

    Sci Am (1896) 74:307

Henry, John E. (? -1892).   Railroad engineer and contrac-
    tor.

    RR Gaz (1892) 24:417

Henry, Joseph (1797-1878).   Electrical engineer; origina-
    tor of practicable method for sending telegraphic
    signals.

    Eng & Min J (1878) 25:340
    Frank Inst J (1880) 110:257-70
    Sci Am (1869) 21:395; (1878) 38:336-7; (1892) 66:193;
        (1899) 81:408-10*

    Appleton's Cyc 3:172-3*
    DAB 8:550-3
    NYT (5/17/1878) 5:6

Henshaw, George Holt (? -1891).   Civil engineer, rail-
    ways and beach fronts.

    Eng Rec (1890-1) 23:104

Herron, Campbell B. (1828-94).   Iron manufacturer.

    AISA Bull (1894) 28:101
    Eng & Min J (1894) 57:444
    Iron Age (1894) 53:897

Herron, Frederick (1852?-97).   Manufacturer of first
    structural steel in U.S.

    Eng & Min J (1897) 63:359
    Eng Rec (1896-7) 35:421

Hessey, D. Stewart (? -1896). Civil engineer, railroads.
   Eng Rec (1896) 34:439

Heusch, Constantine (? -1893). Mining engineer.
   Eng & Min J (1893) 55:468

Hewes, Joseph L. (? -1873). Manufacturer and inventor
   of machinery.
   Sci Am (1873) 29:241

Hewison, Charles William (1830?-96). Mechanical engineer
   and inventor of improvements in marine engines.
   Am Mach (1896) 19:145
   Eng & Min J (1896) 61:91
   Eng Rec (1895-6) 33:129
   Iron Age (1896) 57:310
   Sci Am (1896) 74:67

Hewson, Michael Butt (? -1891). Chief engineer of
   Mississippi & Tennessee Railroad.
   Eng Rec (1890-1) 23:120
   RR Gaz (1891) 23:86

Heywood, Lincoln C. (1868-95). Civil engineer of Lincoln,
   R.I.
   Assoc Eng Soc J (1895) 14:56-7
   Eng Rec (1894-5) 31:129

Hibbert, Stephen Decatur (? -1897). Naval engineer;
   chief engineer of Gulf Squadron.
   Eng Rec (1896-7) 35:333

Hick, William Bayres (1831?-97). Mechanical engineer.
   Eng & Min J (1897) 63:606

Hicks, William Cleveland (1829-85). Civil and mechanical
   engineer; inventor of Hicks engine, improvements in
   fire arms, and sewing machines.
   ASME Trans (1886) 7:828
   Eng News (1885) 14:439

Hidley, Emerson G. (? -1897). Engineer of street railway company.
   Eng Rec (1896-7) 35:355

Higginson, Charles (? -1899). Railroad engineer.
   Am Eng & RR J (1899) 73:213

Hildreth, Russell Wadsworth (1865-95). Mechanical
engineer, specialist in iron construction and bridges.

ASCE Proc (1896) 22:576
ASCE Trans (1896) 36:596-7
ASME Trans (1896) 17:743
Eng Rec (1895-6) 33:57
RR Gaz (1896) 28:30

Hilgard, Julius Erasmus (1825-91). Engineer and geo-
desist, coastal survey.

ASCE Proc (1892) 18:180-1

DAB 9:23

Hill, George A. (1853?-1900). Mechanic, railroads.

Am Mach (1900) 23:728

Hill, Hamilton A. (1832-99). Mechanical engineer.

ASME Trans (1899) 20:1012

Hill, John (? -1898). Civil engineer of Georgia.

Eng Rec (1897-8) 37:183

Hill, Joseph (? -1896). Engineer and operator of Western
railroads.

Eng Rec (1896) 34:325

Hill, Nathaniel Peter (1832-1900). Metallurgist; worked
on smelting process.

Eng & Min J (1900) 69:626,646*

DAB 9:43-4

Hill, Paul (1816?-95). Civil engineer, canals and Hoosac
Tunnel.

Eng Rec (1894-5) 31:309
RR Gaz (1895) 27:206

Hill, Samuel W. (1796?-1889). Civil engineer and geologist.

Eng & Min J (1889) 48:186

Hill, Thomas F. (1805?-94). Iron founder and builder of
machinery.

Eng Rec (1894) 30:134

Hillman, Charles (? -1898). Builder of wooden and iron and steel ships.

Am Mach (1898) 21:984

Himrod, David (? -1877).   Pioneer iron manufacturer in Pa. and Ohio; produced first metal with raw coal.

AISA Bull (1877) 11:316
Iron Age (11/29/1877) 20:11

Hinckley, Francis Edward (1834-1900). Builder of bridges and railroads.

Eng Rec (1900) 42:258
Iron Age (9/13/1900) 66:19

Hines, Dauphine S. (1829-85).   Hydraulic and mechanical engineer.

Am Mach (12/5/1885) 8:8
ASME Trans (1886) 7:828-9
Ind World (12/3/1885) 25:6
Sanit Eng (1885) 12:519

Hippey, A. C. (? -1896). Railroad engineer.

Eng Rec (1896) 34:197
RR Gaz (1896) 28:578,594

Hipply, A. C. (? -1896). Railroad engineer.

Eng Rec (1896) 34:217

Hitchcock, Hiram A. (1857-95). Civil engineer, professor at Dartmouth.

Eng & Min J (1895) 59:83
Eng Rec (1894-5) 31:148
RR Gaz (1895) 27:60

Hitchcock, William J. (1829?-99). Pioneer iron manufacturer of Mahoning Valley.

Am Manuf (1899) 65:450
Eng & Min J (1899) 68:675
Iron Age (11/23/1899) 64:18

Hjortsberg, Maximilian (1825-80). Civil engineer of Chicago.

Assoc Eng Soc J (1881-2) 1:247-9

Hoadley, John Chipman (1818-86). Civil and mechanical
    engineer; inventor of portable engine; manufacturer
    of locomotives and textile machinery.

    Am Mach (11/13/1886) 9:1; (11/27/1886) 9:1-2*
    ASME Trans (1887) 8:724-6
    Sanit Eng (1886) 14:573
    Sci Am (1886) 55:405

    Appleton's Cyc 3:219
    DAB 9:83-4

Hobbs, Alfred Charles (1812-91). Mechanical engineer;
    inventor of Hobbs lock.

    Am Eng & RR J (1891) 65:573
    ASME Trans (1892) 13:263-74
    Sci Am (1891) 65:406*

    DAB 9:95-6

Hobbs, Edwin H. (1835-90). Civil engineer.

    Eng Rec (1890-1) 23:22

Hodge, John (1833-1900). Car builder.

    Am Eng & RR J (1900) 74:388
    RR Gaz (1900) 32:763

Hodgman, George P. (? -1898 or 1899). Locomotive engi-
    neer and designer.

    Am Mach (1899) 22:37
    Iron Age (1/5/1899) 63:20
    RR Gaz (1898) 30:937

Hoe, Richard March (1812-86). Inventor and manufacturer
    of printing presses.

    Am Mach (6/26/1886) 9:7
    ASCE Proc (1890) 16:170-2
    Sci Am (1886) 54:384

    Appleton's Cyc 3:225
    DAB 9:104
    NYT (6/9/1886) 5:4

Hoe, Robert (1814?-84). Manufacturing of printing presses.

    Iron Age (9/18/1884) 34:17
    Sanit Eng (1884) 10:369
    Sci Am (1884) 51:193

    Appleton's Cyc 3:226
    NYT (9/14/1884) 9:4

Hoefer, Eugene (? -1899). Mining engineer of Pacific Coast and South Africa.

AIME Trans (1900) 30:xxxi

Hoepfner, Carl (1857-1900). Metallurgical engineer; inventor in the field of electro-metallurgy.

Eng & Min J (1900) 70:725*
Eng Rec (1900) 42:604

Hoeppner, Arnold J. (? -1895). Civil engineer.

Eng Rec (1895-6) 33:39

Hoff, John B. (1804?-88). Constructor of railroads in Delaware.

Eng Rec (1888) 18:10

Hoffman, E. O. (? -1898). Civil engineer.

Eng Rec (1898) 38:421

Hogan, John J. (? -1896). Inventor of Hogan boiler.

Elec Eng (1896) 22:139
Iron Age (1896) 58:273
Mach (1896-7) 3:24
Power (9/1896) 16:15

Hogan, Peter (1826?-97). Civil and consulting engineer; dams and public works.

Am Eng & RR J (1897) 71:387
Eng & Min J (1897) 64:492
Eng Rec (1897) 36:421

Hoge, William C. (? -1896). Civil engineer.

Eng Rec (1896) 34:343

Hogeland, Justus M. (? -1894). Railroad engineer.

Eng Rec (1894) 30:150

Hogg, Thomas (1808?-81). Master mechanic and railroad engineer of Ohio.

RR Gaz (1881) 13:240

Hogue, Parker P. (? -1899). Inventor of Hogue injector for boilers.

Iron Age (3/30/1899) 63:19

Hoit, David (1831-85). Car builder; superintendent of Gilbert Car Manufacturing Company.

Nat Car (1885) 16:84
RR Gaz (1885) 17:349

Holley, Alexander Lyman (1832-82). Mechanical engineer, responsible for improvements on Bessemer steel process; known as father of American steel manufacture.

AISA Bull (1879) 13:138; (1882) 16:36,309
Am Mach (2/18/1882) 5:8
Am Manuf (2/3/1882) 30:10-1
ASCE Proc (1890) 16:212-5
ASME Trans (1882) 3:29-67; (1883) 4:35
Assoc Eng Soc J (1881-2) 1:241-5
Cassier's (1893-4) 5:233-6*
Eng (1882) 3:57
Eng & Min J (1879) 27:368-9; (1882) 33:61-3*
Eng News (1882) 9:41; (1890) 24:305-6, 315-6*
Ind World (2/16/1882) 18:17*
Iron Age (5/22/1879) 23:14; (2/2/1882) 29:1,20
RR Gaz (1882) 14:77
Sanit Eng (1884-5) 11:168
Sci Am (1882) 46:80*; (1890) 63:224
Van Nostrand's (1882) 26:253-7

Appleton's Cyc 3:235-6*
DAB 9:148-9
NYT (1/30/1882) 5:1

Holley, S. H. (1849-99). Marine engineer, machinist, and inventor.

Am Mach (1899) 22:674
Iron Age (7/13/1899) 64:19

Holloway, Josephus Flavius (1825-96). Mechanical and hydraulic engineer; one of the founders of ASME.

Am Eng & RR J (1896) 70:264
AIME Trans (1896) 26:827-34
Am Mach (1896) 19:867-8*
Am Manuf (1896) 59:447
ASME Trans (1897) 18:612-23
Assoc Eng Soc J (1896) 17:15-6
Cassier's (1896-7) 11:2*, 47-8
Eng (1896) 32:62
Eng & Min J (1896) 62:251
Eng News (1896) 36:173
Eng Rec (1896) 34:253,271
Iron Age (1896) 58:502*
Mach (1896-7) 3:59
RR Gaz (1896) 28:645

DAB 9:155

Holly, Birdsill (? -1894).   Inventor of hydraulic
  machinery.

  Eng Rec (1893-4) 29:360
  Iron Age (1894) 53:851

Hooper, John (? -1889).   Civil engineer.

  Iron Age (1889) 44:1007

Hoopes, Barton (1827?-95).   Manufacturer of bolts, nuts,
  rivets, etc.

  AISA Bull (1895) 29:260
  Am Manuf (1895) 57:771
  Iron Age (1895) 56:1053
  RR Gaz (1895) 27:778

Hoover, D. W. C. (? -1897).   Inventor of photographic
  apparatus.

  Sci Am (1897) 77:218

Hopkins, David A. (1825?-89).   Inventor and manufacturer
  of lead-lined journal bearing.

  Nat Car (1889) 20:189
  RR Gaz (1889) 21:738

Horn, E. B. (? -1872).   Constructor of electro-mechanical
  engines.

  Sci Am (1872) 27:260

Horsey, Thomas Franklin (1841-90).   Civil engineer.

  Eng & Min J (1890) 50:459

Horsford, Eben Norton (1818-93).   Civil engineer, professor at Harvard.

  Eng Rec (1892-3) 27:111

  Appleton's Cyc 3:265-6
  DAB 9:236-7

Horton, Nathan Waller (? -1886).   Mining and mechanical
  engineer.

  Eng & Min J (1886) 41:333,355*
  Sanit Eng (1885-6) 13:543

Hosie, James P. (? -1898).   Civil and mining engineer of
  Pa.

  AIME Trans (1899) 29:xxxi-xxxii

Hoskin, John (1820?-1900). Metallurgist and mechanical engineer.

Am Mach (1900) 23:169

Hotchkiss, Clark Beers (1838?-90). Organizer of district and telegraph message service in N.Y.

Elec Eng (1890) 10:121
Elec World (1890) 16:80

Hotchkiss, Jed (? -1891). Mining engineer, developer of coal and iron in Virginia.

Iron Age (1891) 47:978

Hotchkiss, Jedediah (1828?-99). Military engineer.

AISA Bull (1899) 33:18
RR Gaz (1899) 31:72

Houghton, Hannibal S. (1827-98). Inventor of improvements on sewing machine and envelope making machine.

Am Mach (1898) 21:752

House, Royal Earle (1814-95). Electrical engineer, inventor of electro-magnetic line and printing tele-graph.

Eng Rec (1894-5) 31:237
RR Gaz (1895) 27:175
Sci Am (1888) 59:391*

Appleton's Cyc 3:273
DAB 9:259-60

Houston, David C. (1835-93). Member of Corps of Engi-neers; fortifications and river and harbor improve-ments.

Eng Rec (1892-3) 27:488

Appleton's Cyc 3:273

Houston, John (1828-96). Railroad engineer, Erie Rail-road.

ASCE Trans (1898) 39:694-6
Eng & Min J (1896) 62:227
Eng Rec (1896) 34:253
Iron Age (1896) 58:454
RR Gaz (1896) 28:628

Howard, John W. (1826-92). Builder of power ventilators and wire workers.

Eng Rec (1891-2) 25:190
Iron Age (1892) 49:306

Howard, William B. (? -1898). Building and railroad constructor of Chicago.

Eng Rec (1898) 38:47

Howard, William H. (1797?-1879). Inventor of wire machinery.

Iron Age (9/11/1879) 24:15

Howatt, James P. (1844?-99). Mechanic of Brooklyn Navy Yard.

Am Mach (1899) 22:1119

Howe, John (? -1894). Civil engineer of Providence, R.I.

Eng Rec (1894-5) 31:93

Howell, Charles P. (1848?-99). Naval engineer.

Am Mach (1899) 22:1195

Howell, Charles W. (1859-82). Civil engineer, river works and railroads.

ASCE Proc (1882) 8:120-1

Howell, Courtland D. (1814?-85). Civil engineer.

RR Gaz (1885) 17:557

Howell, George P. (? -1899). Engineer of gravity railways near Scranton, Pa.

Eng Rec (1899) 40:611

Howley, Benjamin Duncan (? -1897). Mechanical engineer and steamboat builder.

Eng Rec (1896-7) 35:465

Hoxey, Thomas Franklin (1841-90). Engineer of reservoirs for Paterson water works.

Am Eng & RR J (1890) 64:522
Eng Rec (1890) 22:308

Hoyt, Carroll Livingston (1866-95). Mechanical engineer.

ASME Trans (1895) 16:1191

Hubbard, Frederick (1817-95). Civil engineer, railroads.

Eng & Min J (1895) 60:423
Eng Rec (1895) 32:399

Hubbard, James G. (? -1894). Master mechanic of New York, Lake Erie, and Western Railroad.

Nat Car (1894) 25:44
RR Gaz (1894) 26:148

Hudson, William S. (1810-81). Mechanical engineer, inventor of improvements in locomotives.

Eng (1881) 2:57
Iron Age (1881) 28:15
RR Gaz (1882) 14:747*
Sci Am (1881) 45:97

DAB 9:342-3
NYT (7/22/1881) 5:6

Hughes, David Edward (1831-1900). Electrical engineer; inventor of printing telegraph and microphone.

Sci Am (1900) 82:67

DAB 9:347-8

Hughes, John (? -1892). Shipbuilder of New Orleans.

Eng Rec (1891-2) 25:174

Hughes, John O. (1832-93). Iron manufacturer.

AISA Bull (1893) 27:317
Eng & Min J (1893) 56:428
Iron Age (1893) 52:759

Hulbert, Thomas H. (1857?-89). Mining engineer.

Eng & Min J (1889) 48:412

Hull, Abram S. (1826-90). Mechanic of railroad shops.

Am Eng & RR J (1890) 64:571
RR Gaz (1890) 22:726

Hull, Daniel (1798-1886). Locomotive engineer.

RR Gaz (1886) 18:220

Humphrey, William Sheldon (1860-95). Civil engineer, railroads in West and Midwest.

ASCE Proc (1895) 21:98

Humphreys, Andrew Atkinson (1810-83). Civil and topographical engineer, U.S. Army; harbor improvements.

ASCE Proc (1890) 16:218-20
RR Gaz (1884) 16:16

DAB 9:371-2
NYT (12/29/1883) 5:5

Hunt, Alfred Ephraim (1855-99). Metallurgist; developer
of method for reducing the aluminum ore.

Am Eng & RR J (1899) 73:213
AIME Trans (1900) 30:xxxii-xxxiii
Am Mach (1899) 22:419
ASME Trans (1899) 20:1012-4
Elec World (1899) 33:601-2
Eng & Min J (1899) 67:527*
Eng News (1899) 41:298-9*
Iron Age (5/4/1899) 63:18*
RR Gaz (1899) 31:322

DAB 9:381-2

Hunt, E. B. (? -1863). Military engineer.

Sci Am (1863) 9:250

NYT (10/2/1863) 4:5

Hunt, Randall (1856-98). Civil and hydraulic engineer,
bridge structures and foundations.

Assoc Eng Soc J (1898) 20:27-30
Eng & Min J (1898) 65:168
Eng Rec (1897-8) 37:205

Hunt, Robert (1807-87). Mining engineer and chemist.

Am Manuf (11/13/1887) 41:13

Hunt, Walter (1796?-1859). Inventor of sewing machine.

Sci Am (1859) 1:21

Hunt, William H. (? -1889). Chief engineer of U.S. Navy.

Am Eng & RR J (1889) 63:388
Eng & Min J (1889) 47:461
RR Gaz (1889) 21:452

Hunter, James (? -1891). Manufacturer of machinery.

Power (7/1891) 11:14

Hunter, Alexander T. (1839-96). Civil engineer and
professor of mathematics.

Eng & Min J (1896) 61:187

Hunter, William (1845?-99). Chief engineer of Central
Railroad in Georgia.

Eng Rec (1899) 40:734
RR Gaz (1899) 31:887

Huntington, Edward (1817?-81). Civil engineer and iron
manufacturer.

RR Gaz (1881) 13:224

Hurst, Benjamin (1810?-90). Locomotive engineer.

RR Gaz (1890) 22:425

Hurt, E. Fletcher (1845-99). Military engineer.

Eng Rec (1899) 40:419

Hussey, C. Curtis (1840?-84). Superintendent of steel
works.

Iron Age (3/6/1884) 33:19

Hussey, Curtis Grubb (1802-93). Manufacturer of crucible
cast steel and sheet copper.

AISA Bull (1892) 26:370
Am Manuf (1893) 52:718
Eng & Min J (1893) 55:396
Ind World (5/11/1893) 40:7

Appleton's Cyc 3:330
DAB 9:430-1

Huston, Charles (1822-97). Iron manufacturer; improver of
mechanical processes in manufacture of steel and iron.

AISA Bull (1897) 31:13
Eng & Min J (1897) 63:71
Iron Age (1/14/1897) 59:19

DAB 9:433-4

Hutchins, C. H. (1858-99). Manufacturer of car roofs.

RR Gaz (1899) 31:619

Hutchins, Carleton B. (1814-94). Manufacturer of
refrigerator cars and freight car roofing.

Nat Car (1894) 25:188
RR Gaz (1894) 26:827

Hyatt, Thaddeus (? -1872). Inventor of glass covered
gratings and other construction improvements.

Sci Am (1872) 26:121

Hyde, William B. (1842?-82). Constructor of milling machinery on west coast; planned water supply for San Francisco.

ASCE Proc (1890) 16:172-4

Hyndman, Edwin K. (1844?-84). Railroad engineer.

RR Gaz (1884) 16:508

Ide, Albert L. (1841-97). Inventor and manufacturer of Ide engine.

ASME Trans (1898) 19:966-7
Cassier's (1898) 14:544-8
Elec Eng (1897) 24:360
Iron Age (10/7/1897) 60:18

Inch, Philip (1836?-98). Chief engineer of U.S. Navy.

Am Mach (1898) 21:811
Eng Rec (1898) 38:443

Ingersoll, Simon (1818-94). Inventor of Ingersoll rock drill.

Eng & Min J (1894) 58:77*
Eng Rec (1894) 30:134

DAB 9:472

Ingraham, Clinton S. (? -1892). Mechanic and machine designer.

Iron Age (1892) 50:291

Ingram, James (1809?-87). Inventor of improvements in sanitary appliances.

Sanit Eng (1887) 16:242

Innes, William P. (? -1893). Railroad engineer, New York & Erie Railroad.

Eng Rec (1893) 28:166

Inshaw, John (? -1896). Inventor of twin screw propeller and injector for filling boilers.

Eng & Min J (1893) 55:157

Inslee, William Harvey (1841-98). Machinist; improver of sewing machine.

Am Mach (1898) 21:587
ASME Trans (1898) 19:980-1

Jackson, John G. (? -1897). Chief engineer of Wilmington & Western Railroad.

Eng Rec (1896-7) 35:333

Jackson, Samuel H. (? -1896). Railroad engineer.

Eng Rec (1896-7) 35:25

Jacobsen, Charles E. A. (? -1897). Sewage and railroad engineer.

Eng Rec (1896-7) 35:245

James, Montgomery (? -1895). Civil engineer; worked on Hoosac Tunnel and bridges on Mississippi River.

Eng Rec (1895-6) 33:75

James, Samuel L (? -1894). Civil engineer.

Eng Rec (1894) 30:150

Janin, Alexis (? -1897). Metallurgist.

Eng & Min J (1897) 63:95

Janney, Morris P. (1850-98). Mechanical engineer.

AIME Trans (1899) 29:xxxii
Eng & Min J (1898) 66:727*

Jauriet, Charles F. (1818?-83). Master mechanic.

RR Gaz (1883) 15:416,447

Jefferies, Edwin (1815-99). Civil engineer and iron manufacturer.

AISA Bull (1899) 33:61
Am Mach (1899) 22:290

Jenkins, David S. (1834-99). Steamboat and railroad engineer.

Iron Age (3/2/1899) 63:22

Jenks, Barton H. (? -1896). Manufacturer of machinery.

Iron Age (1896) 58:1265

Jenks, C. W. (1826-90). Mining engineer, electrician, and manufacturer.

Elec Eng (1890) 9:299

Jennings, Henry C. (1858-94). Civil engineer; specialist in railroad bridges.

ASCE Proc (1894) 20:88

Jennings, Richard (1819-91). Mining engineer and oil producer.

Am Manuf (1891) 49:499

Jernegan, J. L. (? -1881). Mining engineer.

Eng & Min J (1881) 31:246

Jervis, John Bloomfield (1795-1885). Civil engineer, chief engineer of several railroads and canals, including Erie Canal; responsible for major work on Croton Aqueduct.

ASCE Proc (1885) 11:109-16
Eng News (1885) 13:42*; 14:337-8
Iron Age (1/15/1885) 35:17
RR Gaz (1885) 17:40,49-50*
Sanit Eng (1884-5) 11:168
Van Nostrand's (1885) 32:168

Appleton's Cyc 3:430-1*
DAB 10:59-60
NYT (1/14/1885) 2:4

Jervis, William (1813?-86). Railroad engineer.

RR Gaz (1886) 18:633

Jewett, Luther Kendall (? -1894). Inventor of improvements in locomotive construction.

Nat Car (1894) 25:92

Jewett, William B. (1868-94). Civil engineer.

ASCE Proc (1894) 20:202-3

Johns, Henry Ward (1837-98). Manufacturer and inventor; developed commercial uses of asbestos.

Am Mach (1898) 21:133
Elec World (1898) 31:262
Eng & Min J (1898) 65:228
Eng Rec (1897-8) 37:227
Iron Age (2/17/1898) 61:19
RR Gaz (1898) 30:147

Johnson, Charles Roberts (1851-93). Engineer; developer and manufacturer of railroad switches.

Am Eng & RR J (1893) 67:499-500*
ASCE Proc (1894) 20:45-8
ASME Trans (1895) 16:1188-9
Eng & Min J (1893) 56:298
RR Gaz (1893) 25:697

Johnson, Edward H. (? -1896). Consulting and chief engineer of Chicago & Northwestern Railroad.

RR Gaz (1896) 28:99

Johnson, Edwin Ferry (1803-72). Civil engineer, chief engineer of several railroads and canals; inventor of improvements for canal locks, screw power press, and locomotives.

RR Gaz (1872) 4:173
Van Nostrand's (1872) 7:452

DAB 10:96-7
NYT (4/20/1872) 2:4

Johnson, Isaac G. (1831-99). Inventor of projectiles; steel manufacturer.

AISA Bull (1899) 33:100
Am Mach (1899) 22:520
Iron Age (6/8/1899) 63:18
RR Gaz (1899) 31:416
Sci Am (1899) 80:397

Johnson, Samuel F. (1800?-83). Civil engineer, railways.

Assoc Eng Soc J (1884-5) 4:247-8

Johnson, W. J. (? -1898). Civil engineer, copper mines in Arizona.

Eng Rec (1898) 38:179

Johnston, Samuel R. (1833-99). Civil engineer; general roadmaster of Erie Railroad.

Eng Rec (1899) 40:734
RR Gaz (1899) 31:900

Jones, Daniel N. (1829-88). Engineer of machine shops and iron works.

ASME Trans (1889) 10:836

Jones, B. Ellwood (? -1900). Member of engineering corps
of Philadelphia & Reading Railroad.

Eng Rec (1900) 42:330

Jones, David (? -1882). Civil engineer, railroads.

Am Eng & RR J (1882) 55:101

Jones, Griffith (1812-89). Ironmaster and inventor of
improvements in furnaces.

AISA Bull (1889) 23:42

Jones, John (1819?-97) Machinist and manufacturer of
cotton and wool machinery.

Am Mach (1897) 20:440-1
Iron Age (6/3/1897) 59:23

Jones, John H. (1812-89). Inventor in the field of iron
production.

Am Manuf (9/20/1889) 45:11
Eng & Min J (1889) 48:251
Iron Age (1889) 44:449

Jones, L. C. (1830-89). Railroad engineer.

RR Gaz (1889) 21:484

Jones, Wilbur Hodgson (1859-85). Mechanical engineer.

ASME Trans (1886) 7:827

Jones, William Richard (1839-89). Inventor of several
mechanical devices in steelmaking process.

Am Eng & RR J (1889) 63:531
AIME Trans (1889) 18:621
AISA Bull (1889) 23:276*
Am Mach (10/10/1889) 12:6
ASME Trans (1889) 10:838-42
Eng News (1889) 22:351
RR Gaz (1889) 21:652

DAB 10:208-9

Jones, Willis C. (? -1895). Mechanical engineer.

ASME Trans (1895) 16:1198

Jordan, Gabriel (1831-84). Civil engineer; chief engineer
of several railroads.

ASCE Proc (1890) 16:184-5
RR Gaz (1884) 16:826

Joslin, Isaac R. (1842-89). Civil engineer.

Iron Age (1889) 44:646

Judah, Theodore Dehone (1826-63). Railroad engineer; promoter of first railroad across Sierra Nevadas.

ASCE Proc (1897) 23:430-5
ASCE Trans (1897) 38:448-52

DAB 10:229

Judson, Egbert (1812-93). Inventor and manufacturer of explosives; builder of machinery and bridges on West Coast.

Eng Rec (1892-3) 27:130

DAB 10:239

Jungerman, Charles L. (1858?-90). City engineer of Newport, Ky.

Eng Rec (1890) 22:242

Kahler, Charles P. (? -1899). City surveyor in Boston and West.

Eng Rec (1898-9) 39:128

Karnaga, Robert (1844?-1900). Locomotive engineer.

Am Mach (1900) 23:98

Kase, Simon P. (1814-1900). Machinist; inventor of force pump.

AISA Bull (1900) 34:157
Am Mach (1900) 23:872
Iron Age (9/6/1900) 66:21

Keely, John Ernst Worrall (1827-98). Fraudulent inventor of Keely motor.

Am Mach (1898) 21:887-8
Am Manuf (1898) 63:782
Elec Eng (1898) 26:523
Elec World (1898) 32:574
Eng & Min J (1898) 66:644
Eng News (1898) 40:335-6*
RR Gaz (1898) 30:847

Appleton's Cyc 3:499
DAB 10:280-1
NYT (11/19/1898) 7:4

Keen, George V. (1821-97). Manufacturer of tinware.

Iron Age (1/28/1897) 59:21

Keesey, John (? -1898). Mechanical engineer.

Iron Age (12/15/1898) 62:20

Kelley, John F. (1818?-94). Ironmaster; inventor of improvements in steel making.

Eng & Min J (1894) 57:60
Eng Rec (1893-4) 29:102

Kellogg, Charles (? -1891). Civil engineer; inventor of seamless tube.

Am Eng & RR J (1891) 65:187
Am Manuf (1891) 48:239

Kellogg, Montgomery A. (1830-98). Chief engineer of construction of parks in N.Y. City.

Eng Rec (1897-8) 37:447
RR Gaz (1898) 30:299

Kells, Ross (1840-92). Superintendent of motive power.

Am Eng & RR J (1892) 66:192
Nat Car (1892) 23:62

Kelly, William C. (1811-88). Reputed inventor of Bessemer process for making steel.

Am Eng & RR J (1888) 62:191
Am Manuf (2/17/1888) 42:11
Eng & Min J (1888) 45:130
Iron Age (1888) 41:333-4*
Sanit Eng (1887-8) 17:188,220

Appleton's Cyc 3:508-9
DAB 10:311-2

Kelly, Robert (? -1882). Marine engineer.

Eng (1882) 3:130

Kennedy, Archibald M. (? -1897). Bridge builder of Indiana.

Eng Rec (1897) 36:25

Kennedy, John P. (1820?-92). Construction engineer for gas works.

Eng Rec (1891-2) 25:222

123

Kennedy, John S. (? -1892). Mechanical engineer.

Iron Age (1892) 49:415

Kennedy, Leonard W. (? -1898). Inventor; promoter of tunnel in Calif.

Eng Rec (1898) 38:223

Kennedy, Thomas Walter (1823?-96). Iron manufacturer and furnace builder.

AISA Bull (1896) 30:138
Am Manuf (1896) 58:838
Iron Age (1896) 57:1370

Kenyon, A. J. (? -1888). Chief engineer of U.S. Navy.

Eng & Min J (1888) 46:89

Kerrigan, William (? -1890). Railroad engineer.

RR Gaz (1890) 22:548

Kettlewell, Samuel H. (? -1891). Topographical engineer.

Am Eng & RR J (1891) 65:381
Eng Rec (1891) 24:36

Kid, Charlton B. (1836-82). Naval engineer.

Eng News (1882) 9:231

Killebrew, Samuel (1840-99). Civil engineer, railroads in Texas and Mexico.

ASCE Proc (1899) 25:333-6
ASCE Trans (1899) 41:639-42

Kilmer, Irving A. (1858-93). Mechanical engineer; inventor of hay bands, collar for cylinder shafts, etc.

Iron Age (1893) 51:732

Kimball, Alonzo S. (1843-97). Responsible for establishment of departments of physics and electrical engineering at Worcester Polytechnic.

Elec Eng (1897) 24:587
Eng Rec (1897-8) 37:25
Iron Age (12/9/1897) 60:31

Kimball, Edwin A. (? -1898). Inventor and mechanical expert.

Am Mach (1898) 21:888
Eng Rec (1898) 38:531

Kimball, Hiram (1845-99). Inventor of machinery for
   turnbuckle manufacture.

   AIME Trans (1900) 30:xxxiv
   Am Mach (1899) 22:244
   ASME Trans (1899) 20:1011
   Iron Age (3/16/1899) 63:19

Kimball, James M. (? -1898). Civil engineer, railroads.

   Eng Rec (1897-8) 37:249

King, John Newton (? -1886). Master car builder of
   Chesapeake & Ohio Railroad.

   RR Gaz (1886) 18:170

King, Louis (? -1888). Inventor; chief engineer of
   Etna Iron Works.

   Am Manuf (5/11/1888) 42:11

King, William R. (? -1898). Member of U.S. Corps of
   Engineers, river harbor and fortifications.

   Eng Rec (1897-8) 37:535
   RR Gaz (1898) 30:381

King, Zenas (1818-92). Manufacturer of iron bridges and
   boilers.

   Am Eng & RR J (1892) 66:581
   Eng Rec (1892) 26:338

Kingsbury, Horace G. (? -1897). City engineer of
   Newark, Ohio.

   Eng Rec (1896-7) 35:487

Kingsley, William C. (1833-85). Civil engineer and contractor; connected with Brooklyn Bridge and water works
   in Brooklyn.

   ASCE Proc (1892) 22:700-3
   ASCE Trans (1896) 36:612-5
   RR Gaz (1885) 17:143

   Appleton's Cyc 3:549
   NYT (2/21/1885) 1:7

Kirby, James P. (? -1898). Civil engineer.

   Eng Rec (1897-8) 37:447

Kirkpatrick, David Warden (1857-97). Civil engineer.

AISA Bull (1897) 31:285
Eng & Min J (1897) 64:762
Eng Rec (1897-8) 37:25
Iron Age (12/9/1897) 60:31

Kirkwood, James P. (1807-77). Civil engineer, several
railroads and water works; chief engineer of Brooklyn.

ASCE Proc (1878) 4:60-5

Appleton's Cyc 3:557

Klohs, Abraham (1819-85). Machinist and master mechanic.

RR Gaz (1885) 17:270

Kloman, Andrew (1827-80). Inventor of saw and straighten-
ing machinery; improvements in manufacture of rolled
iron shapes.

AISA Bull (1881) 15:82
ASCE Proc (1881) 7:122-4
Iron Age (1/27/1881) 27:1*

Knabb, John G. (1808-98). Railroad engineer.

Am Mach (1898) 21:568

Kneas, Strickland (1821-84). Civil engineer, municipal
works and railroads.

RR Gaz (1884) 16:41

Appleton's Cyc 3:561
DAB 10:455-6

Knight, Jacob Brown (1833-79). Mechanical engineer.

Eng News (1879) 6:82
Frank Inst J (1879) 107:280-1

Knight, William B. (1848-90). Civil engineer, railways,
cable railways, and sewers.

Am Eng & RR J (1891) 65:43
ASCE Proc (1891) 17:277-8
Eng Rec (1890-1) 23:22
RR Gaz (1890) 22:902

Koch, Edward Cabot (1859-98). Mining engineer in West.

AIME Trans (1899) 29:xxxii

Koenig, Frederick (? -1899). Constructing engineer for
   railroads in Pittsburgh area.

   Eng Rec (1899) 40:658

Koerner, F. (? -1900). Mining and mechanical engineer.

   Eng & Min J (1900) 70:736

Kraemer, James Madison (? -1900). Civil engineer of
   Philadelphia.

   Iron Age (4/26/1900) 65:21

Krause, Harrison Y. (1838?-97). Manufacturer of agricultural implements.

   Iron Age (4/1/1897) 59:15

Krause, Herman F. (? -1897). Topographical engineer.

   Eng Rec (1897) 36:377

Kretsinger, William H. (1816-94). Manufacturer of
   farming implements.

   Iron Age (1894) 53:166

Kruesi, John (1843-99). Mechanical engineer, inventor
   of underground tubing system; chief mechanical engineer
   of General Electric.

   Am Mach (1899) 22:177
   ASME Trans (1899) 20:1009-10
   Elec Eng (1899) 27:260*
   Elec World (1899) 33:290*
   Eng & Min J (1899) 67:270
   Eng Rec (1898-9) 39:315
   Iron Age (3/2/1899) 63:22
   RR Gaz (1899) 31:160
   Sci Am (1899) 80:163

   DAB 10:509-10

Kustel, Guido (1817?-82). Metallurgist.

   Eng & Min J (1882) 34:145

Kyan, John Howard (1775-1850). Inventor of kyanizing
   process for preserving wood.

   Am Eng & RR J (1850) 23:24

   Appleton's Cyc 3:579

Labram, George (1860?-1900). Mechanical and consulting
engineer in South Africa.

Elec World (1900) 35:348

Laing, John (1817?-97). Chief engineer of Union Railroad
Company.

Eng & Min J (1897) 64:702
Eng Rec (1897-8) 37:3
RR Gaz (1897) 29:859

Lake, George Burt (1844-84). Civil engineer.

Assoc Eng Soc J (1883-4) 3:212-3
RR Gaz (1884) 16:347

Lamb, W. B. (? -1889). Civil engineer.

RR Gaz (1889) 21:844

Lamborn, Robert H. (? -1895). Railway engineer.

Eng Rec (1894-5) 31:129

Lander, James (1838-94). Superintendent of motive power
and master mechanic of New York, New Haven & Hartford
Railroad.

Am Eng & RR J (1894) 68:472

Lane, James C. (1823-88). Civil engineer, mineralogist,
and surveyor.

Am Eng & RR J (1889) 63:47
Eng & Min J (1888) 46:506
Eng Rec (1888-9) 19:54
RR Gaz (1888) 20:843

Lane, Moses (1823-82). Hydraulic engineer.

ASCE Proc (1893) 19:58-9
Assoc Eng Soc J (1881-2) 1:249-50
Eng News (1882) 9:36; (1890) 23:46*
RR Gaz (1882) 14:77

Lane, Philander P. (1821-99). Machinist and mechanic of
Cincinnati.

Am Manuf (1899) 65:509
Eng (1900) 37:13-4
Eng Rec (1899) 40:682

Lange, A. W. (? -1888). Civil engineer.

Am Eng & RR J (1888) 62:572

128

LaNoue, G. Morgan (? -1896). City engineer of Houston.

Eng Rec (1896) 34:21

Lanphear, O. A. (? -1894). Mechanical engineer.

ASME Trans (1894) 15:1192

Lapham, William G. (1817?-73). Civil engineer, railroads.

RR Gaz (1873) 5:445

Large, Augustus (1835?-1900). Locomotive engineer, Pennsylvania Railroad.

Am Mach (1900) 23:436

Larimer, Joseph (1851?-94). Inventor and constructor of special shapes in iron and steel.

Ind World (8/30/1894) 43:6

Larkin, J. Edward (? -1891). Bridge and railroad engineer.

RR Gaz (1891) 23:275

Larrabee, Charles S. (? -1894). Inventor of fuse for naval purposes.

Eng & Min J (1894) 57:12

Latimer, Charles (1827-88). Railroad engineer, chief engineer of railroads in N.Y., Pa., and Ohio; inventor of rerailing guard and system of naval signals.

ASCE Proc (1889) 15:137-40
Assoc Eng Soc J (1888) 7:201-7, 475-7
Eng & Min J (1888) 45:238
Eng News (1888) 19:267-8*
RR Gaz (1888) 20:212
Sanit Eng (1887-8) 17:263

Appleton's Cyc 3:625

Latrobe, Benjamin H. (1806-78). Civil engineer, railroads; consulting engineer for Hoosac Tunnel.

Eng News (1878) 5:343
RR Gaz (1878) 10:513

Appleton's Cyc 3:627
DAB 11:26

Latrobe, John Hazlehurst Boneval (1803-91). Engineer and inventor, railroads and bridges.

Elec Eng (1891) 12:391*
Sci Am (1889) 60:214*

Appleton's Cyc 3:627*
DAB 11:27-8

Lauder, James Nelson (1847?-94). Master mechanic and superintendent of motive power for New York, New Haven & Hartford Railroad.

Eng Rec (1894) 30:214
Nat Car (1894) 25:156*
RR Gaz (1894) 26:603,619

Laurie, James (1811-75). First president of ASCE; railroad and bridge engineer; consulting engineer for Hoosac Tunnel.

ASCE Proc (1897) 23:168-70
ASCE Trans (1897) 37:553

Appleton's Cyc 3:632
DAB 11:36-7

Lauth, Bernard (1820-94). Inventor of cold-rolled shafting and mill appliances.

AISA Bull (1894) 28:140
Am Manuf (1894) 55:17
Iron Age (1894) 54:16

Laweon, Peter B. (1810-79). Mechanical and steam engineer.

Sci Am (1879) 40:394

Lawrence, James S. (1811-86). Surveyor and dredging operator; constructor of forts.

Assoc Eng Soc J (1885-6) 5:390-1

Lawton, Elbridge (1826-89). Chief engineer of U.S. Navy.

RR Gaz (1889) 21:500

Lay, John Louis (1832-99). Inventor of Lay torpedo for submarines.

Am Mach (1899) 22:369

Appleton's Cyc 3:644
DAB 11:64-5

Lea, Albert Miller (1807?-91). Member of engineering corps.

Eng Rec (1890-1) 23:121

Lea, Robert (? -1895). Machine builder and foundryman.

Iron Age (1895) 55:1335

Lean, George R. (? -1897). Inventor, mechanic, and electrician.

Elec World (1897) 30:262

Learned, Edward (1820?-86). Civil engineer and contractor, railroads.

RR Gaz (1886) 18:152

LeConte, Llewellyn (? -1900). Professor of civil engineering.

RR Gaz (1900) 32:783

Ledlie, James Hewett (1832-82). Civil engineer, canals and railroads.

RR Gaz (1882) 14:513

Appleton's Cyc 3:654
NYT (8/16/1882) 5:3

Lee, Alexander Young (1838?-97). Civil engineer and architect.

Am Manuf (1897) 61:624
Eng Rec (1897) 36:465

Lee, Francis D. (? -1885). Civil engineer, inventor of marine torpedo.

Sanit Eng (1885) 12:274

Lee, Leighton (? -1898). Mechanical engineer.

Am Mach (1898) 21:888
Eng Rec (1898) 38:531

Lee, Richard Henry (1831-91). Railroad and harbor engineer; chief engineer for several railroads in South.

Eng & Min J (1892) 53:308

Lee, Stephen States (1812?-92). Railroad constructor, coal mine operator, and civil engineer.

Am Eng & RR J (1892) 66:479
Eng & Min J (1892) 54:204
Eng Rec (1892) 26:194
RR Gaz (1892) 24:642

Lee, Thomas J. (1806?-91). Civil engineer.

Eng Rec (1891-2) 25:106

Leers, Frank A. (1844-90). Civil and mechanical engineer, elevated railroads, mines, and foundries.

Am Eng & RR J (1890) 64:331
ASCE Proc (1894) 20:85
Eng Rec (1889-90) 21:386
RR Gaz (1890) 22:367

Lefevre, Henry (1841?-99). Developer of mining industry in Colombia.

Eng & Min J (1899) 68:74

Lehman, William R. (? -1890). Engineer in charge of Union Pacific surveying party.

Eng Rec (1890-1) 23:2

Leighton, James T. (1837?-92). Master car builder.

Nat Car (1892) 23:158

Leighton, Thomas (? -1886). Builder of bridges for railroads.

RR Gaz (1886) 18:101
Sanit Eng (1885-6) 13:256

Leisenring, John (1819-84). Civil engineer, canals and mining.

Am Manuf (8/29/1884) 35:18
Iron Age (8/28/1884) 34:17
RR Gaz (1884) 16:641

Leitch, Robert R. (1850?-99). Chief engineer of U.S. Navy.

Am Mach (1899) 22:244

Lenhardt, George (? -1897). Electrical and mechanical engineer.

Eng Rec (1897-8) 37:3

Leonard, M. B. (1856-97). Electrical engineer; inventor of railway signal apparatus.

Elec Eng (1897) 23:192
Elec World (1897) 29:248

Lessig, Samuel (1823?-98). Inventor and manufacturer of agricultural implements.

Am Mach (1898) 21:511
Iron Age (6/30/1898) 61:18

Lester, Charles H. (? -1899). Military engineer.

Eng Rec (1899) 40:538

Lester, John Henry (1816?-1900). Inventor and manufacturer of sewing machines.

Am Mach (1900) 23:71

Leuffer, George W. (1814?-99). Railroad engineer.

Eng Rec (1899) 40:182

Levrat, Henri (1828-91). Superintending and constructing engineer.

Eng (1891) 21:145*

Lewis, George T. (1817-1900). Inventor of commercial processes for the refinement of industrial cotton seed oil.

Eng & Min J (1900) 69:116
Iron Age (1/25/1900) 65:29

Lewis, Isaac C. (1812-93). Manufacturer of metal goods.

Iron Age (1893) 52:1130

Lewis, Isaiah William Penn (1808-56). Civil engineer, bridges and lighthouses.

ASCE Proc (1897) 23:429-30
ASCE Trans (1897) 38:453-4

Lewis, Joseph (? -1894). Inventor of Lewis locomotive valve, principles, etc.

Iron Age (1894) 54:956
RR Gaz (1894) 26:812

Lewis, Richard Bernard (1825-79). Civil engineer, railroads.

Eng News (1879) 6:385
RR Gaz (1879) 11:611

Lewis, Samuel (?-1898). Inventor of multiple drill used in removing Diamond Reef and Flood Rock in N.Y. Harbor.

Eng Rec (1897-8) 37:469

133

Lewis, Thomas S. (1840?-97). Hydraulic engineer.
Eng & Min J (1897) 63:215

Libby, Forrest L. (1864-94). Civil engineer of Boston.
Assoc Eng Soc J (1896) 17:24-5

Lienau, D. B. (? -1890). Mining engineer.
Eng & Min J (1890) 49:645

Lighthall, William A. (? -1881). Steam engineer; designer and builder of marine engines.
Sci Am (1881) 44:49

Lincoln, George S. (1819?-94). Developer of Lincoln milling machine; head of Phoenix Iron Works.
Am Mach (4/19/1894) 17:10

Lindroth, C. O. (1853?-1900). Mechanical engineer.
Am Mach (1900) 23:1032

Linn, W. H. (? -1895). City engineer of Springfield, Ohio; canals and railroads in West.
Eng Rec (1895) 32:416

Lipe, Charles E. (1851-95). Inventor of broom making machinery; developer of milling machinery.
ASME Trans (1895) 16:1195

Livingood, Henry S. (? -1892). Mining engineer.
Eng & Min J (1892) 54:276

Lobdell, George G. (1817-94). Manufacturer of car wheels.
AISA Bull (1894) 28:53
Eng & Min J (1894) 57:228
Iron Age (1894) 53:464
Nat Car (1894) 25:60
RR Gaz (1894) 26:183

Locke, Augustus Woodbury (1846-93). Civil engineer, railroads and tunnels; worked on Hoosac Tunnel.
ASCE Proc (1893) 19:172-4
Assoc Eng Soc J (1894) 13:10-1
Eng & Min J (1893) 55:498
Eng Rec (1892-3) 27:488
RR Gaz (1893) 25:384

Locke, Sylvanus Dyer (1833-96). Civil and mechanical engineer; developer of harvesting machinery.

ASME Trans (1897) 18:1094-100
Eng Rec (1896) 34:325

Lockett, Samuel H. (? -1891). Waterworks engineer in Calif.

Am Eng & RR J (1891) 65:574
Eng Rec (1891) 24:374
RR Gaz (1891) 23:790

Lockwood, John (1814-91). Gas and hydraulic engineer.

ASCE Proc (1892) 18:192-3
Eng Rec (1891-2) 25:2

Lockwood, Joseph A. (? -1897). Civil engineer.

Eng Rec (1897) 36:267

Loiseau, Emile Francois (1831-86). Inventor of practical method for making compressed artificial fuel from coal dust.

ASME Trans (1886) 7:830-1
Frank Inst J (1886) 122:304
Sci Am (1886) 54:358

Lombaert, Herman J. (1815-85). Civil engineer, canals and railroads.

RR Gaz (1885) 17:173

NYT (3/11/1885) 5:6

Loper, R. F. (? -1880). Inventor and shipbuilder.

Sci Am (1880) 43:344

NYT (11/9/1880) 3:5

Lord, George D. (? -1896). Engineer and contractor, railroads and canals.

Eng Rec (1895-6) 33:183

Lord, Horace (1815-85). Mechanical engineer.

ASME Trans (1885) 6:872

Lord, Russell Farnham (1838?-99). Civil and mining engineer, Delaware & Hudson Canal Company.

AIME Trans (1900) 30:xxxiv
Eng & Min J (1899) 68:104
Eng Rec (1899) 40:156
RR Gaz (1899) 31:530

Lorenz, William (1826-84). Civil engineer; chief engineer of Philadelphia & Reading Railroad.

Iron Age (1/8/1885) 35:17
RR Gaz (1885) 17:14

NYT (12/30/1884) 5:5

Lotz, William Herman (1838-94). Mechanical engineer; inventor of heating and ventilating apparatus.

ASCE Proc (1894) 20:74-6

Loughridge, William (? -1890). Inventor of brakes.

Am Eng & RR J (1890) 64:237

Louis, Henry (? -1895). Mining engineer.

Eng & Min J (1895) 60:399

Lovell, Melvin N. (1845?-95). Manufacturer of dynamos, motors, wringers, and household specialties.

Iron Age (1895) 56:1104

Lovett, Thomas David (1823-97). Railroad and bridge engineer in Ohio and Midwest.

ASCE Proc (1898) 24:712-8
ASCE Trans (1898) 40:571-7

Low, Gorham P. (? -1894). Civil and hydraulic engineer, railroads and dams.

ASCE Proc (1894) 20:72-3
Eng Rec (1893-4) 29:118

Low, Joel Bacon (1824?-98). Mining engineer.

Eng & Min J (1898) 65:678

Low, Sigismund (1825-98). Civil engineer, early railroad work in Pa.

Eng & Min J (1898) 65:228
Eng Rec (1897-8) 37:227
RR Gaz (1898) 30:169

Lowry, Joseph L. (? -1893). Mechanical engineer; inventor of pumping engine.

Eng Rec (1892-3) 27:231

Lowthrop, Francis C. (1810-90). Civil engineer, designer of early iron railroad bridges; inventor of railroad turntable.

Am Eng & RR J (1890) 64:331
ASCE Proc (1894) 20:196-8
Eng Rec (1890) 22:2

Lucas, George L. (? -1891). Railroad engineer.

Eng Rec (1891) 24:118

Ludden, Henry D. (1837-99). Civil engineer; city engineer of Detroit.

Eng Rec (1899) 40:300

Ludlam, Joseph S. (1837-96). Miner, mechanic, and marine engineer.

ASME Trans (1897) 18:1091-2

Ludlow, R. C. (1826-95). Machinist; founder of wire company in St. Louis.

Iron Age (1895) 56:72

Lufkin, Harvey Lamb (1857-96). Electrician; responsible for installing electric power in rolling mills.

AIEE Trans (1896) 13:444-5
Elec Eng (1896) 22:653*
Elec World (1897) 29:47*
Iron Age (1896) 58:1265

Lundberg, Orlof Raynor Gottfried (? -1892). Mechanical engineer and superintendent of iron works.

Eng Rec (1891-2) 25:290

Lunt, Clarence Williams (1850-84). Civil engineer.

Assoc Eng Soc J (1886-7) 6:229-31

Lush, William (1850?-95). Railroad engineer.

RR Gaz (1895) 27:29

Luther, Henry M. (1855?-92). Engineer of Philadelphia & Reading Coal & Iron Company.

Eng Rec (1891-2) 25:342

Lyman, Azel Storrs (1815-85). Inventor of fountain pens,
    alarms, air engine, refrigerating car, etc.

    Sci Am (1885) 53:145

Lyman, Gad (1810?-85). Locomotive engineer.

    Am Eng & RR J (1884-5) 58:105-6
    Eng (1885) 10:1

Lyon, Lewis (? -1891). Responsible for introduction of
    cable system to N.Y.

    Am Eng & RR J (1891) 65:573

Lyon, William M. (1809-89). Iron manufacturer of Pitts-
    burgh.

    AISA Bull (1889) 23:195
    ASCE Proc (1890) 16:112-3
    Iron Age (1889) 44:55

Lyte, Francis Asbury (1854-96). Surveyor and engineer,
    railroads and cities.

    ASCE Proc (1897) 23:471
    ASCE Trans (1897) 38:461

McAlpine, Charles LeGrand (1828-84). Civil engineer,
    canals and railroads, N.Y. State.

    ASCE Proc (1897) 23:326
    ASCE Trans (1897) 37:563
    RR Gaz (1884) 16:55

    NYT (1/12/1884) 5:3

McAlpine, William Jarvis (1812-90). Civil engineer,
    canals, hydraulic works, railroads, bridges and
    parks in N.Y. State and City.

    Am Eng & RR J (1890) 64:140
    ASCE Proc (1892) 18:115-23
    Eng & Min J (1890) 49:230,309*
    Eng News (1890) 23:182,223-5*
    Eng Rec (1889-90) 21:177-8
    Iron Age (1890) 45:301
    RR Gaz (1890) 22:134
    Sci Am (1890) 62:130

    Appleton's Cyc 4:71
    DAB 11:548-9

McArthur, James (1832-93). Civil engineer and contractor, river improvements and railroads.

Eng & Min J (1893) 55:227
Eng Rec (1892-3) 27:292
RR Gaz (1893) 25:216

McBride, William J. (? -1894). Mechanical engineer; inventor of coal-hoisting machinery.

Eng Rec (1894) 30:166

McCallum, Daniel Craig (1815-78). Civil engineer, military railroads; inventor of McCallum arched truss bridge.

RR Gaz (1879) 11:9
Sci Am (1879) 40:37

Appleton's Cyc 4:76
DAB 11:565-6
NYT (12/24/1878) 2:7

McCammon, William (1811?-81). Civil engineer.

RR Gaz (1881) 13:200

McClellan, Carswell (1835-92). Civil and military engineer, constructor works and western railroads; U.S. Civil Asst. Engineer.

Am Eng & RR J (1892) 66:192
Eng Rec (1891-2) 25:242
RR Gaz (1892) 24:197

McClellan, George Brinton (1826-85). Military engineer, railroads in Midwest; river and harbor work in Texas.

Eng & Min J (1885) 40:302
Iron Age (11/5/1885) 36:21
RR Gaz (1885) 17:717

Appleton's Cyc 4:78-84*
DAB 11:581-5

McClure, Robert John (1841?-99). Civil engineer, railroads; consultant.

RR Gaz (1899) 31:216
West Soc Eng J (1900) 5:427-9*

MacCord, Charles W. (1822?-98). Mechanical engineer and
technical journalist; associate editor of Power.

Am Mach (1898) 21:454-5
Eng Rec (1898) 38:25
Power (9/1898) 18:19
RR Gaz (1898) 30:419

McCormick, Cyrus H. (1809-84). Inventor of reaper
and self-sharpening horizontal plow; iron
manufacturer.

Am Mach (6/7/1884) 7:2
Ind World (5/15/1884) 22:7
Iron Age (5/15/1884) 33:34
Sci Am (1869) 21:408; (1884) 50:321

Appleton's Cyc 4:94-5*
DAB 11:607-9

McCormick, Leander J. (1819-1900). Manufacturer of
harvesting machinery.

Am Mach (1900) 23:192
Iron Age (2/22/1900) 65:21
RR Gaz (1900) 32:144
Sci Am (1900) 82:154

DAB 11:610

McCrum, John S. (1838?-99). Superintendent of motive
power, Kansas City, Fort Scott & Memphis Railroad.

Am Mach (1899) 22:290

McCullough, James W. (1827-97). Engineer, railroad
and blast furnace work.

Eng & Min J (1897) 58:486

McDonald, James M. (? -1895). Civil engineer and
bridge builder; iron manufacturer.

Eng Rec (1894-5) 31:111

McDonald, Marshall (? -1895). Professor of geology and
engineering at Virginia Military Institute; U.S.
Fisheries Commissioner.

Eng Rec (1895) 32:255

McDonald, Ronald Trevor (1849-98). Manufacturer of lighting apparatus; involved in electrical railway business.

Am Eng & RR J (1899) 73:63
Elec Eng (1898) 26:653
Elec World (1898) 32:734*

McDonough, J. W. (? -1897). Electrician and inventor.

Elec Eng (1897) 24:144

McDonough, Thomas (1822?-93). Marine and railroad engineer.

Eng (1893) 25:124
Eng & Min J (1894) 57:611

McElroy, Samuel (1825-98). Versatile civil and mechanical engineer; best known for hydraulic work and as chief engineer, Continental Railroad.

ASME Trans (1899) 20:1004
Eng News (1881) 8:433
Eng Rec (1899) 39:83
Iron Age (12/15/1898) 62:20
RR Gaz (1898) 30:902
West Soc Eng J (1899) 4:237-9

McEntee, James S. (1800?-87). Civil engineer, Erie and Union Canals; railroad surveyor.

Am Eng & RR J (1887) 61:348
RR Gaz (1887) 19:460

McEwan, Henry D. (1839-94). Naval engineer.

Iron Age (1894) 54:709

McFarland, John (1826-93). Locomotive engineer and machinist; superintendent of motive power, Chesapeake & Ohio Railroad.

Am Mach (8/11/1883) 6:5
RR Gaz (1883) 15:486,498

McFarland, Walter M. (1836?-88). Assistant engineer, U.S. Navy.

Am Eng & RR J (1888) 62:429
Eng Rec (1888) 18:106
RR Gaz (1888) 20:496

MacFarlane, James (1819-85). Mining surveyor.

Eng & Min J (1885) 40:286
RR Gaz (1885) 17:702

McIntire, C. (? -1890). Inventor of McIntire joint and connector.

Elec World (1890) 15:52

McKay, Donald (1810-80). Naval architect and master shipbuilder; builder of marine and locomotive engines.

Sci Am (1880) 43:228-9

Appleton's Cyc 4:126-7
DAB 12:72-3
NYT (9/22/1880) 5:5

McKay, Hugh (? -1898). Planner and operator of first steam-propelled boat, Erie Canal.

Eng Rec (1897-8) 37:557

McKee, James Harper (1818?-95). Car manufacturer.

Iron Age (1895) 56:1002

Mackenzie, Philip Wallace (1824-91). Inventor of Mackenzie blower and cupola for smelting iron ore.

AISA Bull (1891) 25:173

Mackenzie, Ronald Slidell (1840-89). Soldier and military engineer.

Eng Rec (1888-9) 19:114
DAB 12:95-6

MacKinney, William C. (1848-95). Mechanical engineer.

ASME Trans (1896) 17:738-9

MacLean, F. F. Grosvenor (1854?-91). Metallurgist and professor at Johns Hopkins.

Iron Age (1891) 47:543

McLeod, George (? -1877). Railroad engineer.

RR Gaz (1877) 9:221,233

MacLeod, John (? -1900). Civil engineer, railroads.

Eng Rec (1900) 41:114
RR Gaz (1900) 32:79

McMillan, James A. (? -1897). Civil engineer, railroads.

Eng Rec (1896-7) 35:179

McMillan, Joseph (1832?-99). Bridge engineer.

Iron Age (2/2/1899) 63:26

McMillan, William (1831-99). Civil engineer, parks.

Eng Rec (1899) 40:254

McMurtrie, John (1848-99). Surveyor and railroad
engineer.

Eng Rec (1899) 39:290

McQueen, Walter (1816?-93). Mechanical engineer and
inventor of improvements in locomotive construction.

Am Eng & RR J (1893) 67:354*
Eng & Min J (1893) 55:588
Eng Rec (1892-3) 27:52
Nat Car (1893) 24:112
RR Gaz (1893) 25:473

Macy, Arthur (1852?-91). Mining engineer.

ASCE Proc (1896) 22:617
ASCE Trans (1897) 37:562-3

Maddock, W. H. (? -1896). Mechanical engineer,
Pittsburgh.

Eng Rec (1896-7) 35:25

Maguire, Edward (1846?-92). Military engineer.

Am Eng & RR J (1892) 66:528
Eng & Min J (1892) 54:372
Eng Rec (1892) 26:306

Mahan, Dennis Hart (1802-71). Professor of civil and
military engineering, West Point; author of text-
books.

ASCE Proc (1893) 19:161-2
Sci Am (1871) 25:213

Appleton's Cyc 4:176
DAB 12:209-10

Mahone, William (1826-95). Railroad engineer and
executive, South.

Eng Rec (1895) 32:345
RR Gaz (1895) 27:680

Appleton's Cyc 4:177
DAB 12:211-2
NYT (10/9/1895) 13:4

Mahony, James (1828-92). Master mechanic and inventor of mechanical devices.

ASME Trans (1892) 13:681-2

Maillefert, Gustave Jaques (1823-97). Mechanical engineer.

ASME Trans (1898) 19:975

Main, Thomas (1828?-96). Mechanical engineer; designer of steam engines for ships.

Eng & Min J (1896) 61:547
Eng Rec (1896) 33:453
Iron Age (1896) 57:1315

Mallory, W. H. (1840-82). Inventor of screw steering propeller.

Sci Am (1882) 47:359

Mann, George Edward (1845-97). Civil engineer, city works and railroads, Buffalo, N.Y.

Eng Rec (1897) 36:399
RR Gaz (1897) 29:754

Mann, George H. (? -1874). Civil engineer, canals, bridges, and channels.

Van Nostrand's (1875) 12:96

Mann, John G. (1834-99). Chief engineer, Mobile & Ohio Railroad.

RR Gaz (1899) 31:358

Mapes, John L. (? -1891). Civil engineer, railroads and bridges.

Eng Rec (1891) 24:396

Marsh, James S. (1821-89). Manufacturer of pig iron and agricultural implements; inventor of self-raking reaping machine.

AISA Bull (1889) 23:165

Marsh, Sylvester (1803-84). Inventor of machinery for mechanical handling of grain; builder of Mt. Washington Railroad, N.H.

RR Gaz (1885) 17:14

Appleton's Cyc 4:219
DAB 12:303-4

144

Marsh, William W. (1827?-92). Iron manufacturer.

Eng & Min J (1892) 54:228
Eng Rec (1892) 26:210

Marshall, Charles Alfred (1855-89). Civil engineer, bridges; engineer of tests, Cambria Iron Company.

ASCE Proc (1890) 16:99-100

Marshall, George (? -1896). Signal engineer and inventor.

Eng Rec (1896) 34:289

Marshall, Moses (1812-87). Inventor of knitting machine.

Am Manuf (9/16/1887) 41:11

Marshall, Robert Emory (1862-90). Railroad engineer; superintendent of motive power, Philadelphia, Wilmington & Baltimore Railroad.

ASME Trans (1897) 18:1100

Marsland, Edward (1829?-98). Mechanical engineer; designer of steamships and dredging machinery.

Am Mach (1898) 21:492
Eng (1892) 22:99*
Eng Rec (1898) 38:91
Iron Age (6/30/1898) 61:17

Martin, Edward (1810-93). Civil engineer, railroads.

Eng & Min J (1893) 56:598
Eng Rec (1893-4) 29:34
RR Gaz (1893) 25:896

Martin, Edward H. (1832-92). Civil engineer and machine designer; inventor of horse-rake and Martin pump.

Am Eng & RR J (1893) 67:101
Eng Rec (1892-3) 27:111
Iron Age (1/5/1893) 51:24

Martin, Robert Kirkwood (1835-93). Chief engineer, Baltimore Water Department.

ASCE Proc (1894) 20:43-5
Eng Rec (1893-4) 29:2

Marvin, Walter K. (1824-85). Manufacturer of Marvin safe; inventor of locks and bolts.

Am Manuf (12/25/1885) 37:11

145

Mason, Eddy D. (1832-75). Engineer, railroads and bridges.

ASCE Proc (1873-5) 1:329-30
Eng News (1875) 2:22*

Mason, Henry (1829?-98). Mechanical engineer; supervised engine installation for warships.

Am Mach (1898) 21:679
Eng & Min J (1898) 66:284
Iron Age (9/1/1898) 62:17

Mason, Marcus (? -1898). Mechanical engineer and inventor.

Am Mach (1898) 21:587
Iron Age (8/4/1898) 62:19

Mason, Roswell B. (1805?-92). Civil engineer, railroads and canals.

Am Eng & RR J (1892) 66:96
Assoc Eng Soc J (1893) 12:433-6*
Eng Rec (1891-2) 25:90

Mason, William (1808-83). Manufacturer of machinery and locomotives; inventor of "self acting mule" for spinning cotton and other fibers.

Am Manuf (6/8/1883) 32:9
RR Gaz (1883) 15:336,341-2*
Sci Am (1883) 48:336

DAB 12:377-8

Massey, Albert P. (1842-98). Mechanical engineer; inventor of vacuum brakes.

RR Gaz (1898) 30:501

Maxon, Thomas (1837?-1900). Mechanical engineer and inventor; patentee of lever jack, endless chain, etc.

Am Manuf (1900) 66:47
Iron Age (2/1/1900) 65:21

Maxwell, Eugene Lafelle (1851?-95). Manufacturer of machinery.

Am Eng & RR J (1895) 69:139
Mach (3/1894-5) 1:7

May, Frank A. (1854-83?). Civil engineer.

Assoc Eng Soc J (1884) 3:209

Mayall, T. J. (? -1888). Prolific inventor, worked on electric railroads and underground conduits for wires.

Elec World (1888) 11:133

Maynard, Edward (1813-91). Dentist; inventor of firearms and dental instruments.

Sci Am (1891) 64:304

Appleton's Cyc 4:276
DAB 12:457-8

Meiggs, Henry (1811-77). Builder of South American railroads.

Eng & Min J (1878) 26:437*
Eng News (1877) 4:289
Iron Age (10/18/1877) 20:15

Appleton's Cyc 4:287
DAB 12:501-2
NYT (10/12/1877) 5:4

Meigs, Montgomery Cunningham (1816-92). Military and civil engineer, buildings and public water supply system in Washington, D. C.

Am Eng & RR J (1892) 66:96
Eng Rec (1891-2) 25:90
Sci Am (1892) 66:71-2*

Appleton's Cyc 4:289*
DAB 12:507-8

Mein, Thomas (1838-1900). Mining engineer, California, Alaska and South Africa.

Eng & Min J (1900) 69:554

Menge, Joseph (? -1894). Civil engineer; inventor of dredge boat.

Eng & Min J (1894) 57:420

Menger, Leslie G. (? -1896). Civil engineer and surveyor, New Jersey.

Eng Rec (1895-6) 33:380

Mercur, Frederick (1837-88). Mining engineer; superintendent of Lehigh Valley Coal Co.; official of Lehigh Valley Railroad.

Am Eng & RR J (1888) 62:94
ASCE Proc (1891) 17:157-60
Eng & Min J (1888) 45:31
Eng Rec (1887-8) 17:109
Ind World (1/19/1888) 30:24
RR Gaz (1/13/1888) 20:29

Mercur, James (? -1896). Professor of civil and military engineering.

Eng Rec (1896) 33:363

Mergenthaler, Ottmar (1854-99). Mechanical engineer; improver of linotype machine.

Am Mach (1899) 22:1093-5*
Sci Am (1899) 81:314*

DAB 12:549-50

Merrill, William E. (1837?-91). Member, Corps of Engineers; supervised river and harbor improvements, canalization of portion of Ohio River.

Am Eng & RR J (1892) 66:48
ASCE Proc (1892) 18:90-3
Eng Rec (1891-2) 25:38
RR Gaz (1891) 23:906

Appleton's Cyc 4:306
DAB 12:568-9

Merritt, Thomas G. (? -1889). Engineer, bridges.

Am Eng & RR J (1889) 63:196

Merry, Frederick C. (? -1900). Architect and engineer.

Eng Rec (1900) 41:233

Meryweather, Thomas Arthur (? -1897). Civil engineer, railroads in Philadelphia.

Eng Rec (1896-7) 35:311
RR Gaz (1897) 29:212

Messimer, Hillary (1833-98). Mechanical engineer.

ASME Trans (1899) 20:1004-5
Eng Rec (1898) 39:105

Meston, A. W. (1867?-93). Electrical engineer.

Elec Eng (1893) 15:510-1

Meyer, George F. (? -1894). Mechanical engineer and
    inventor; consulting engineer for refrigerating
    machine company.

    Eng Rec (1894-5) 31:3
    Iron Age (1894) 54:95-6

Meyer, J. G. Arnold (1841-1900). Draftsman; associate
    editor of American Machinist.

    Am Mach (1900) 23:726*
    ASME Trans (1900) 21:1167-8

Michaelis, Otto E. (1844?-90). Military engineer,
    experimented with electrical devices.

    Am Eng & RR J (1890) 64:281
    ASCE Proc (1891) 17:132-7
    Elec Eng (1890) 9:330
    RR Gaz (1890) 22:327

Michie, William Roberts (1866-99). Civil engineer,
    railroads and tunnels.

    ASCE Proc (1899) 25:276-7
    ASCE Trans (1899) 41:647
    RR Gaz (1899) 31:107

Mickley, Edwin (1830?-98). Mining engineer.

    AISA Bull (1898) 32:109
    Am Mach (1898) 21:511
    Eng & Min J (1898) 66:14
    Eng Rec (1898) 38:91
    Iron Age (7/21/1898) 62:19

Mifflin, Joseph (? -1885). Civil engineer, railroads.

    RR Gaz (1885) 17:143

Mifflin, Samuel W. (? -1885). Civil engineer, rail-
    roads.

    RR Gaz (1885) 17:541

Milholland, John B. (? -1896). Manufacturer of engines
    and machinery.

    Iron Age (1896) 57:1140

Millar, William (? -1895). Civil engineer.

    Eng Rec (1894-5) 31:237

Miller, Edward (? -1872). Railroad engineer, Little
    Schuylkill & Susquehanna Railroad.

    RR Gaz (1872) 4:78

Miller, Ezra (1812-85). Civil engineer; inventor of
    Miller car coupler and buffer.

    Nat Car (1885) 16:109
    RR Gaz (1885) 17:461

    Appleton's Cyc 4:325
    DAB 12:624-5

Miller, George Clinton (? -1896). Civil engineer;
    importer of tiles; dealer in earthenware and
    plumbing supplies.

    Eng Rec (1895-6) 33:255

Miller, Horace B. (1840-98). A founder of the American
    Machinist.

    Am Eng & RR J (1898) 72:418
    Am Mach (1898) 21:829
    ASME Trans (1899) 20:1000-1
    Elec Eng (1898) 26:498
    RR Gaz (1898) 30:837

Miller, Lewis (1829-99). Manufacturer of agricultural
    machinery, Ohio.

    Am Mach (1899) 22:158
    ASME Trans (1899) 20:1004-16
    Iron Age (3/2/1899) 63:21

    DAB 12:634-5

Miller, Samuel H. (1829-91). Civil engineer, railroads
    in Midwest.

    ASCE Proc (1891) 17:208-10

Miller, Silvanus (1851-97). Civil engineer; important
    to development of railroads in Central and South
    America.

    ASCE Trans (1898) 39:696
    Eng & Min J (1897) 64:762
    Eng Rec (1897-8) 37:69

Miller, Sydney Greene (1816?-1900). Civil engineer.

    Eng Rec (1900) 42:629

Miller, William (1820-88). Manufacturer of iron and military equipment.

Am Eng & RR J (1888) 62:525
Am Manuf (9/28/1888) 43:11
ASME Trans (1889) 10:833
Iron Age (1888) 42:468

Millholland, John H. (? -1897). Railroad engineer.

Eng Rec (1897) 36:69

Mills, Edgar (1827-93). Civil engineer; surveyor for Hudson River Railroad; gold miner in California.

Eng & Min J (1893) 55:36
Eng Rec (1892-3) 27:130

Mills, Emory W. (1842?-82). Improver of the steam engine.

Am Mach (9/9/1882) 5:5

Milner, John Turner (1826-98). Civil engineer; manufacturer of pig iron; developer of railroads in South.

Eng & Min J (1898) 66:254
RR Gaz (1898) 30:617

DAB 13:19-20

Minott, Hiram P. (? -1891). Proprietor of machine works.

ASME Trans (1892) 13:675

Moen, Philip L. (1824-91). Wire manufacturer.

Am Manuf (1891) 48:359
Elec Eng (1891) 11:506
Iron Age (1891) 47:834

Moffatt, Edward Stewart (1844-93). Metallurgist; manufacturer of steel rails; president of Lackawanna Iron & Steel Company.

AISA Bull (1893) 27:236
ASME Trans (1893) 14:1449
Eng & Min J (1893) 56:170
Iron Age (1893) 52:269

Moister, Isaac R. (? -1896). Mining engineer, Lehigh Valley Coal Company.

Eng Rec (1896-7) 35:69

Moler, J. Douglas (? -1891). Civil engineer.

Eng Rec (1891) 24:412

Molyneaux, John (1812?-98). Machinist; builder of gunboats during Civil War.

Iron Age (2/24/1898) 61:29

Monroe, James H. (1820-96). Constructor of and authority on steam boilers.

Iron Age (1896) 58:1019

Monroe, John Albert (? -1891). Civil engineer, specialized in foundations for bridges.

Am Eng & RR J (1891) 65:332
ASCE Proc (1893) 19:148-51
Eng & Min J (1891) 51:700
Eng Rec (1891) 24:2
Iron Age (1891) 47:1171
RR Gaz (1891) 23:438

Montague, Samuel S. (1837-83). Chief engineer, Central Pacific Railroad.

Eng News (1883) 10:476,478
RR Gaz (1883) 15:661

Montgomery, James (? -1889). Mechanical engineer and inventor; developer of boilers.

Am Mach (1/17/1885) 8:5

Mooney, William (? -1891). Gas engineer and architect.

Eng Rec (1890-1) 23:120

Moore, Charles Carroll (? -1895). Civil and mechanical engineer.

Eng Rec (1894-5) 31:417

Moore, Gilpin (1841-1900). Inventor of agricultural implements; developed Gilpin plow.

Iron Age (2/22/1900) 65:20

Moore, Henry C. (1813-89). Civil engineer, railroads and canals.

Assoc Eng Soc J (1889) 8:492-4

Moore, James (1813?-97). Civil engineer, New Jersey
   railroads.

   Eng Rec (1897) 36:245
   RR Gaz (1897) 29:595

Moore, James O. (? -1894). Civil engineer,
   & Ohio Railroad.

   RR Gaz (1894) 26;130

Moore, Samuel (1803?-94). Iron manufacturer; master
   mechanic, Central Railroad of New Jersey.

   Am Mach (4/5/1894) 17:8
   Eng Rec (1893-4) 29:264
   Nat Car (1894) 25:60
   RR Gaz (1894) 26:220

Moorhead, James Kennedy (1806-84). Builder of canals;
   pioneer in commercial telegraphy.

   AISA Bull (1884) 18:67
   Iron Age (3/13/1884) 33:17

   Appleton's Cyc 4:385
   DAB 13:147-8

Moorhead, Joel Barlow (1813-89). Railroad and canal
   builder; iron manufacturer.

   AISA Bull (1889) 23:300
   Am Manuf (11/1/1889) 45:11
   Eng & Min J (1889) 48:39
   Eng Rec (1889) 20:314
   Iron Age (1889) 44:685

Moran, Paul (? -1898). Machinist; organizer of
   machine and shipbuilding firm.

   Am Mach (1898) 21:829

Mordecai, Alfred (1804-87). Military engineer;
   assistant engineer, Mexico & Pacific Railroad.

   Am Eng & RR J (1887) 61:557

   Appleton's Cyc 4:389
   DAB 13:153-4

Morgan, James (1843-92). Superintendent of American
   Iron and Steel Company.

   AISA Bull (1892) 26:189
   ASME Trans (1893) 14:1442
   Eng Rec (1892) 26:54

Morgan, Thomas R. (1834-97). Mechanical engineer; inventor and manufacturer of steam hammers.

Am Manuf (1897) 61:371
ASME Trans (1898) 19:968
Eng & Min J (1897) 64:312
Eng Rec (1897) 36:311
Iron Age (9/9/1897) 60:13

Morley, Isaac (1823?-99). City engineer of Pittsburgh.

Eng & Min J (1899) 68:374

Morley, James Henry (1824?-89). Civil engineer, railroads.

ASCE Proc (1890) 16:110-1
Eng & Min J (1889) 48:228

Morley, William Raymond (1846-83). Civil engineer, Mexican Central Railroad.

ASCE Proc (1883) 9:121-6
RR Gaz (1883) 15:14,721-2

Morrell, Daniel Johnson (1821-85). Prominent iron and steel manufacturer, associated with Cambria Iron Company.

AISA Bull (1885) 19:228
Eng & Min J (1885) 40:123
Ind World (8/27/1885) 25:6
Iron Age (8/27/1885) 36:1*
RR Gaz (1885) 17:557
Sanit Eng (1885) 12:269

Morris, Gouverneur (1813-88). Builder of canals and railroads.

RR Gaz (1888) 20:564

Morris, Gouverneur (1847-97). Civil engineer, worked on Croton Aqueduct and several railroads; participated in coal mining operations in Northwest.

ASCE Trans (1898) 39:698-9
Assoc Eng Soc J (1898) 20:61-2
Eng & Min J (1898) 65:48
Eng Rec (1897-8) 37:117
Iron Age (1/6/1898) 65:21
RR Gaz (1898) 30:16

Morris, Robert C. (1817-92). Chief engineer,
    Nashville, Chattanooga & St. Louis Railroad.

    Am Eng & RR J (1892) 66:581
    ASCE Proc (1892) 18:220-2
    RR Gaz (1892) 24:870

Morris, Thomas B. (1843?-85). Civil engineer.

    RR Gaz (1885) 17:779

Morris, Walter J. (1836?-83). Railroad engineer,
    N.Y. State, Midwest, South America and Balkans.

    Iron Age (5/24/1883) 31:13

Morrison, James (? -1894). Railroad engineer, South.

    Eng Rec (1894) 30:248

Morrison, R. McKay (1843-98). Marine engineer.

    Am Mach (1898) 21:415

Morse, Cyrus Bullard (? -1896). Mechanical engineer
    and inventor of improvements in metalworking,
    woodworking and textile machinery.

    Eng & Min J (1896) 61:211
    Eng Rec (1895-6) 33:219

Morse, James Otis (1818-83). Civil engineer, canals
    and railways; responsible for construction of
    large buildings in N.Y. City.

    ASCE Proc (1893) 19:47

Morse, Samuel Finley Breese (1791-1872). Inventor of
    Morse code and telegraph.

    Frank Inst J (1882) 93-374-5; (1882) 94:360
    Sci Am (1869) 21:409; (1872) 26:247,258*

    Appleton's Cyc 4:424-8*
    DAB 13:247-51
    NYT (4/3/1872) 5:2

Morse, Sidney Edward (1794-1871). Inventor of
    engraving process and of bathometer; journalist.

    Sci Am (1872) 26:21

    Appleton's Cyc 4:428
    DAB 13:251-2
    NYT (12/24/1871) 3:5

Morse, Stephen A. (1826?-98). Inventor of twist dril
    Am Mach (1898) 21:984
    Eng Rec (1898-9) 39:105
    Iron Age (1/5/1899) 63:20*

Morton, John Henry (1847-92). Railroad engineer,
Colorado.
    ASCE Proc (1892) 18:105-6
    RR Gaz (1892) 24:145

Mott, Henry A. (1832-96). Chemist and engineer.
    Eng & Min J (1896) 62:467
    Sci Am (1896) 75:375

Mott, Jordan L. (1798-1867?). Inventor of coal-
burning furnaces.
    Sci Am (1869) 21:395

Moulton, J. B. (1810-97). Civil engineer, canals and
railroads.
    Eng Rec (1896-7) 35:179
    RR Gaz (1897) 29:86

Mowbray, George Mordey (1814-91). Improver of
manufacturing methods for nitroglycerine;
introduced its use in blasting the Hoosac Tunnel.
    Am Eng & RR J (1891) 65:381
    Eng Rec (1891) 24:52
    DAB 13:297-8

Mowton, Charles C. (1833-89). Gas engineer.
    Eng Rec (1889) 20:263

Muhlenberg, Edward D. (1832?-83). Civil engineer,
railroads.
    RR Gaz (1883) 15:172
    NYT (3/12/1883) 4:7

Muller, Maurice A. (? -1899). Draftsman and
mechanical engineer.
    ASME Trans (1899) 20:1016

Mulligan, John (1820?-98). Locomotive engineer;
president of Connecticut River Railroad.
    RR Gaz (1898) 30:147

Mullin, John W. (? -1895). Civil and mining engineer.

Eng Rec (1895) 32:380

Mullin, Joseph P. (1854-1900). Mechanical engineer.

Am Mach (1900) 23:146
ASME Trans (1900) 21:1164

Munroe, James H. (? -1896). Constructor of boilers and steam engines.

Eng Rec (1896) 34:457

Munzer, William (? -1884). Builder of steam engines.

Am Mach (10/11/1884) 7:5

Murphy, J. H. (? -1897). Civil engineer.

Eng Rec (1897-8) 37:139

Murphy, John G. (? -1900). Mining engineer.

Eng & Min J (1900) 69:356
Iron Age (3/22/1900) 65:19

Murphy, John McLeod (1827-71). Civil engineer; studied manufacture and use of petroleum and asphalt.

Eng & Min J (1871) 11:378

Appleton's Cyc 4:466

Murphy, John W. (1828-74). Civil engineer; worked on Erie canal, bridges in Pa.

Frank Inst J (1874) 98:302-10
Van Nostrand's (1875) 12:76

DAB 13:355

Murray, James (1812?-95). Engineer, iron works and machine shops.

RR Gaz (1895) 27:227

Myers, Henry M. (1831-98). Inventor of shovel-making machinery; manufacturer of shovels.

Am Manuf (1898) 63:350

Mynderse, Edward (1818?-96). Manufacturer of pumps.

Eng Rec (1896-7) 35:91
Iron Age (1/14/1897) 59:20

Napier, Arthur Howell (1861?-95). Sanitary engineer; specialized in plumbing and drainage of buildings.

Eng Rec (1895-6) 33:57

Neafie, Jacob G. (1815?-98). Shipbuilder and engine builder.

AISA Bull (1898) 32:20
Iron Age (1/20/1898) 61:21

Neal, John (? -1900). Civil engineer.

Eng Rec (1900) 42:66

Neale, Deodatus Hilin (1849-93). Mechanical engineer.

Am Eng & RR J (1893) 67:254
Eng & Min J (1893) 55:324
Eng Rec (1892-3) 27:393
Nat Car (1893) 24:78

Needham, William L. (? -1877). Locomotive engineer; inventor of train signal.

RR Gaz (1877) 9:319

Neilson, Robert (1837-96). Railroad engineer; general superintendent of Philadelphia & Erie Railroad.

Am Eng & RR J (1896) 70:306
ASCE Proc (1897) 23:127-8
ASCE Trans (1897) 37:564-5
Eng Rec (1896) 34:361
RR Gaz (1896) 28:727

Nelson, Peter N. (1822-99). Machinist and engineer; developed paper feeder and ice harvesting machinery.

Am Mach (1899) 22:851
Iron Age (9/7/1899) 64:17

Nes, Charles M. (? -1896). Developed Nes silicon method of converting iron ore into steel.

Eng Rec (1896) 34:41

Nevons, Hiram (1836-94). Civil engineer; superintendent of Cambridge Water Board.

Assoc Eng Soc J (1895) 14:55-6
Eng Rec (1894) 30:3

Newberry, Wolcott E. (1863?-98). Mining engineer.

Eng & Min J (1898) 65:768

Newcombe, George Loring (1811-99). Manufacturer
  of steam engines and boilers.

  Iron Age (3/12/1899) 63:22

Newell, James S. (1824-99). Mechanical engineer
  and manufacturer of special machinery.

  Iron Age (5/25/1899) 63:27

Newell, John (1830-94). Railroad engineer; general
  manager and president of several railroads.

  Am Eng & RR J (1894) 68:472
  ASCE Proc (1894) 20:172-3
  Eng Rec (1894) 30:214
  Nat Car (1894) 25:140
  RR Gaz (1894) 26:603,618*

Newell, Richard (? -1894). Superintendent and chief
  engineer of Midland Terminal road in Colorado.

  Eng Rec (1894-5) 31:75
  RR Gaz (1894) 26:893

Newton, Isaac (1837-84). Civil engineer; vessel
  inspector; worked on rapid transit development
  and public water system, N.Y. City.

  ASCE Proc (1885) 11:128-9
  Iron Age (10/2/1884) 34:17
  RR Gaz (1884) 16:725
  Sanit Eng (1884) 10:409-10

  Appleton's Cyc 4:507

Newton, John (1823-95). Military engineer and soldier.

  Cassier's (1893-4) 5:233-6*
  Eng & Min J (1895) 59:419
  Eng Rec (1894-5) 31:399
  Pop Sci (1895) 47:288

  Appleton's Cyc 4:508-9*
  DAB 13:473-4
  NYT (5/2/1895) 3:6

Newton, W. H. (? -1889). Inventor of sand pump and
  winged float for deepening and clearing river
  channels.

  Eng Rec (1889) 20:156

Nichols, Edward (1850-92). Civil and mechanical engineer; employed in locomotive works, Brooklyn.

Am Eng & RR J (1892) 66:96
AIME Trans (1892-3) 21:76-9
AISA Bull (1892) 26:13
Iron Age (1892) 49:70
Nat Car (1892) 23:31
RR Gaz (1892) 24:49

Nichols, Norman James (1852-94). Engineer, canals and waterworks in western U.S. and South America.

ASCE Proc (1896) 22:154
ASCE Trans (1896) 36:559

Nichols, William Ripley (1847-86). Sanitary engineer; expert in water analysis; noted for public health research.

Assoc Eng Soc J (1886-7) 6:100-5
Eng & Min J (1886) 42:117

Appleton's Cyc 4:514

Nicholson, William Thomas (1834-93). Founder and president of Nicholson File Co.; inventor of file-cutting machine.

ASME Trans (1893) 14:1451
Iron Age (1893) 52:773-4*

DAB 13:509

Nickerson, Louis (? -1877). Civil engineer.

RR Gaz (1877) 9:469

Nickerson, Thomas (1810-92). Railway builder.

Eng Rec (1892) 26:136
RR Gaz (1892) 24:569

Nicoll, William Leonard (? -1887). Naval engineer.

ASME Trans (1887) 8:728

Nicolls, G. A. (1817-96). Civil engineer; president of Philadelphia and Reading Railroad.

RR Gaz (1886) 18:422

Nicolson, Samuel (1791?-1868). Inventor; improved steering apparatus for vessels.

Sci Am (1868) 18:57

Appleton's Cyc 4:518

Nimick, Alexander (1820-98). Iron and steel manufacturer.

AISA Bull (1899) 33:4
Am Mach (1898) 21:984
Am Manuf (1898) 63:917
Eng & Min J (1898) 66:764

Nishwitz, Walter (1830?-1900). Manufacturer of
agricultural implements; invented mowing machine.

Am Mach (1900) 23:305
Iron Age (3/22/1900) 65:19

Nitze, Henry B. C. (1867-1900). Mining engineer.

Eng & Min J (1900) 69:646*

Noble, Charles E. (1824?-99). Railroad engineer.

Eng Rec (1899) 40:711

Noble, Samuel (1834-88). Ironmaster; manufacturer of
pig iron, South.

AISA Bull (1888) 22:257
Eng & Min J (1888) 46:133
Ind World (8/23/1888) 31:5
Iron Age (1888) 42:320*
RR Gaz (1888) 20:564

DAB 13:540

Noble, William (1828?-98). Mining engineer.

Eng & Min J (1898) 66:734
Eng Rec (1898) 37:249
Iron Age (2/17/1898) 61:19

Nock, Edward (? -1883). Pioneer of American puddling;
introduced system into Pittsburgh steel mills.

Am Manuf (4/27/1883) 32:11

NYT (4/23/1883) 1:4

Nock, Thomas Gill (1827-90). Locomotive builder;
president, New York Locomotive Works.

Am Eng & RR J (1890) 64:281
AISA Bull (1890) 24:125
Nat Car (1890) 21:74
RR Gaz (1890) 22:312

Norris, George (? -1896). Civil engineer, railroads
in Maine and Vermont.

Eng Rec (1896) 33:399

Northup, Nicholas Augustus (1849?-96). Civil engineer, Brooklyn city works.

Eng & Min J (1896) 61:211
Eng Rec (1895-6) 33:219

Norton, Frederick O. (1838-92). Cement manufacturer and expert.

ASCE Proc (1892) 18:219-20
Eng Rec (1892) 26:290

Nostitz, Hans von (1836?-91). Engineer and draftsman.

Eng Rec (1890) 23:22

Nostrand, John Lott (1829?-1900). Civil engineer and surveyor; chief engineer of Brooklyn elevated railway.

Eng Rec (1900) 42:403
RR Gaz (1900) 32:711

Nott, Samuel (1815-99). Civil engineer, surveyor and superintendent of railroads and waterworks.

Eng Rec (1899) 40:442
RR Gaz (1899) 31:700

Nourse, Edwin Green (1849-97). Railroad engineer; constructor of several railway terminals.

ASCE Trans (1898) 39:699
Eng Rec (1897-8) 37:25
RR Gaz (1897) 29:897
West Soc Eng J (1898) 2:848-50*

Noyes, Albert Franklin (1850-96). Civil engineer; authority on waterworks and sewerage.

ASCE Proc (1896) 22:607-10
ASCE Trans (1896) 36:560-3
Assoc Eng Soc J (1897) 18:10-4
Eng & Min J (1896) 62:371
Eng Rec (1896) 34:361
RR Gaz (1896) 28:727

Nugent, George E. (? -1896). City engineer, Cincinnati; railroad engineer.

Eng Rec (1896) 34:101

Nutt, Henry Clay (1833?-92). Railroad engineer;
   president of Atlantic & Pacific, and other railroads.

   Am Eng & RR J (1892) 66:431
   Eng Rec (1892) 26:180
   Nat Car (1892) 23:142
   RR Gaz (1892) 24:624

Nystrom, John William (1824-85). Civil engineer; water
   supply problems, Philadelphia.

   Am Mach (6/27/1885) 8:6
   RR Gaz (1885) 17:318

O'Brien, John Z. (? -1898). Civil engineer,
   railroads.

   Eng Rec (1898) 39:105

O'Brien, Robert E. (1833-96). Railway engineer.

   Eng Rec (1896) 34:379
   RR Gaz (1896) 28:746

O'Connell, John C. (1837-98). Naval engineer.

   ASME Trans (1898) 19:982-3

Odenweller, Anthony (1821?-99). Military engineer.

   Eng Rec (1899) 40:491

O'Donavan, John (1814?-94). Locomotive engineer,
   South America.

   RR Gaz (1894) 26:893

Ogden, William Butler (1805-77). Civil engineer,
   canals and railroads near Chicago; president,
   Union Pacific Railway Company.

   ASCE Proc (1878) 4:67-71

   Appleton's Cyc 4:562
   DAB 8:644-5

Olmstead, Charles (? -1899). Electrical manufacturer.

   Elec World (1899) 34:920

Onderdonk, John Remsen (1840-88). Engineer, San
   Francisco sea wall and other harbor and dock works.

   Am Eng & RR J (1889) 63:47
   Iron Age (1888) 42:822

Oothout, E. Austin (1861?-94). Electrical engineer.
   Eng & Min J (1894) 57:132

Orford, John M. (? -1897). Electrical engineer.
   Elec Eng (1897) 23:351

Orton, William (1826-78). President of Western Union
   Telegraph Co.; promoted research and standardization
   in telegraphic engineering.
   Sci Am (1878) 38:309

   Appleton's Cyc 4:596
   DAB 14:65-6

Ortton, John (1825-93). Mechanical engineer, railroads.
   Am Eng & RR J (1893) 67:149

Osborne, Richard Boyse (1815-99). Railroad engineer
   and inventor.
   Eng Rec (1899) 40:658
   Iron Age (12/7/1899) 64:19
   RR Gaz (1899) 31:870

Osgood, Robert R. (? -1898). Inventor; patentee
   of single dipper dredge.
   Am Mach (1898) 21:587

Otis, Charles G. (1831-93). Elevator manufacturer.
   Eng & Min J (1893) 56:170
   Eng Rec (1893) 28:166
   Iron Age (1893) 52:269

Otley, James Walsh (? -1888). Engineer, railroads.
   Am Eng & RR J (1888) 62:94

Oviatt, James S. (1845-90). Civil engineer;
   responsible for city sewer system, Cleveland.
   Assoc Eng Soc J (1890) 9:324

Owen, Edward H. (? -1890). Mechanical engineer;
   performed experimental testing of stationary
   steam and pumping engines.
   ASME Trans (1890) 11:1154

Packard, Loren (1843?-95). Master car builder, New
York Central Railroad.

Am Eng & RR J (1895) 69:139-40
ASME Trans (1895) 16:1192-3
Nat Car (1895) 26:42
RR Gaz (1895) 27:127

Packer, Asa (1805-79). Constructor of canal locks on
upper Lehigh River; builder of Lehigh Valley Railroad; founder of Lehigh University.

Eng & Min J (1879) 27:367
Iron Age (5/29/1879) 23:7
RR Gaz (1879) 11:287-8
Sci Am (1879) 40:374

Appleton's Cyc 4:620-1
DAB 14:131-2
NYT (5/18/1879) 2:5

Packer, Thomas B. (? -1896). Mechanical engineer.

Eng Rec (1896) 34:343

Paddock, George H. (? -1899). Civil engineer.

Eng Rec (1899) 40:254

Paddock, Joseph Hill (1856-94). Mechanical engineer,
iron and coke works; chief engineer, H. C. Frick
Coke Company.

ASCE Proc (1894) 20:89-90
Eng & Min J (1894) 57:324
Eng Rec (1893-4) 29:296

Padgham, Frank W. (1864-91). Draftsman and engineer.

ASME Trans (1891) 12:1055

Page, Charles Grafton (1812-68). Physicist and
experimenter in electricity; invented electro-
magnetic engine.

Sci Am (1865) 13:400

Appleton's Cyc 4:623
DAB 14:135-6

Page, Ezekiel (1811?-73). Inventor of woodworking
machine for turning oars.

Sci Am (1873) 29:113

Page, John (1816?-90). Civil engineer; worked on Erie and St. Lawrence Canals.

Eng Rec (1890) 22:66

Paige, William H. (1841?-85). Inventor of steel-tired car wheel.

RR Gaz (1885) 17:748

Paine, William H. (1828-90). Civil engineer; constructing engineer for East River Bridge, New York City.

Am Eng & RR J (1891) 65:91
ASCE Proc (1891) 17:160-3
Eng & Min J (1891) 51:69
Iron Age (1891) 47:63
RR Gaz (1891) 23:34
Sci Am (1891) 64:17

Palmer, Harry (? -1894). Civil engineer, Surveying Department, Wilmington, Delaware.

Eng Rec (1893-4) 29:392

Palmer, S. (1822?-94). Superintendent of construction, Western Union Telegraph Company.

Nat Car (1895) 26:10

Pankhurst, John F. (1838-98). Vice President of Globe Iron Works; responsible for development of ship-building and marine engineering industry, Midwest.

Am Mach (1898) 21:435
Eng & Min J (1898) 65:708
Iron Age (6/9/1898) 61:19-20

Pardee, Ario (1810-92). Civil engineer and anthracite coal operator.

AISA Bull (1892) 26:90
Eng Rec (1891-2) 25:290
Iron Age (1892) 49:729-30
RR Gaz (1892) 24:255

Appleton's Cyc 4:644
DAB 14:200-1

Park, James, Jr. (1820-83). Leading iron and steel
   manufacturer, Pittsburgh; introduced Siemens
   gas furnace to U.S.

   AISA Bull (1883) 17:116
   Eng & Min J (1883) 35:233
   Ind World (4/26/1883) 24:5
   Iron Age (4/26/1883) 31:15
   RR Gaz (1883) 15:271

   Appleton's Cyc 4:648
   DAB 14:205-7

Parke, John Grubb (1827-1900). Soldier and military
   engineer.

   Eng Rec (1900) 42:629

   Appleton's Cyc 4:649*
   DAB 14:211-2

Parke, Lyman Curtis (1834-99). Mining engineer and
   manufacturer of mining machinery, Pacific coast.

   Eng & Min J (1899) 68:634*

Parker, Charles Eddy (1842-97). Hardware manufacturer;
   noted for innovative manufacturing methods.

   Iron Age (5/20/1897) 59:42*

Parker, Charles H. (? -1897). Designer and builder
   of mining machinery, power plants, textile and
   stove machinery, and bridges.

   ASME Trans (1898) 19:965-6

Parker, Ely Samuel (1828-95). Civil engineer and soldier.

   Eng Rec (1895) 32:255

   Appleton's Cyc 4:650
   DAB 14:219-20

Parker, George Alanson (1822-87). Civil engineer,
   railroads and bridges.

   Assoc Eng Soc J (1889) 8:334-9
   Sanit Eng (1886-7) 15:578

Parrish, George H. (1819-98). Mechanical engineer
   and anthracite operator.

   Eng & Min J (1898) 66:734

Parrott, Peter Pearse (1811-96). Iron manufacturer.

AISA Bull (1896) 30:181
Eng & Min J (1896) 62:131
Eng Rec (1896) 34:177

Parry, Charles T. (1821-87). Proprietor of Baldwin
locomotive works; introduced piecework system to
locomotive manufacture.

Am Eng & RR J (1887) 61:346
Nat Cat (1886) 17:101

DAB 14:262-3

Parsons, Calvin W. (? -1899). Mechanical engineer.

Iron Age (1/5/1899) 63:20

Parsons, Charles C. (? -1894). Mining engineer.

Eng Rec (1894) 30:320

Parsons, Charles O. (1847?-94). Mining engineer.

Eng & Min J (1894) 58:388

Parsons, Henry C. (1839?-94). Railroad engineer,
C & O Railroad.

Eng & Min J (1894) 58:11

Parsons, Henry C. (1833-89). Machinist; foreman
and partner in machine shop.

Am Mach (7/4/1889) 12:8
ASME Trans (1889) 10:838

Patten, Frank Jarvis (1852?-1900). Electrical
engineer; invented system of multiplex telegraphy.

Elec Eng (1890) 9:157*
Elec World (1900) 36:793,798*
Eng Rec (1900) 42:476

Patton, William Henry (1831?-92). Mining and
mechanical engineer; superintendent of silver
mines, Nebraska.

ASME Trans (1893) 14:1442; (1894) 15:1194-5
Eng & Min J (1892) 54:157
Eng Rec (1892) 26:180

Paul, George B. (1837?-1900). Civil engineer;
developed steam canal boat.

Elec World (1900) 35:237

Peacock, George (1823?-1899). Proprietor of Peacock Iron Works, Selma, Alabama; inventor of Peacock self-oiling wheel.

Am Mach (1899) 22:341
Iron Age (4/13/1899) 63:18

Pedale, Charles (? -1893). Railroad mechanic and superintendent.

Am Eng & RR J (1893) 67:254

Peirce, Eldridge (1827-75). Machinist and locomotive engineer.

RR Gaz (1875) 7:295

Pelham, F. B. (1865?-95). Civil engineer, Michigan Central Railroad.

RR Gaz (1895) 27:126

Pendleton, John Montgomery (1844?-1900). Inventor of storage batteries and other electric devices.

Elec World (1900) 36:304

Percival, George Sydney (1867-92). Mechanical engineer.

Am Mach (8/18/1892) 15:8
ASME Trans (1892) 13:682-3
Eng & Min J (1892) 54:132
Eng Rec (1892) 26:150
Sci Am (1892) 67:113

Perkins, Jacob (1766-1849). Inventor of steam gun, high pressure steam boiler, and other mechanical devices.

Appleton's Mech Mag (1851) 1:693-5
Sci Am (1848-9) 4:405

Appleton's Cyc 4:729
DAB 14:472-3

Perry, Nelson W. (1853-98). Electrical engineer; writer for technical journals.

Elec Eng (1898) 25:347
Elec World (1898) 31:430
Eng & Min J (1898) 65:438
Eng Rec (1897-8) 37:381

Peters, Richard (1810-89). Civil engineer, railroads.

RR Gaz (1889) 21:133

DAB 14:510-1

Peters, Samuel (1847-99). Chemist and construction engineer, several iron and steel works.

AIME Trans (1900) 30:xxxvii

Petsch, Julius D. (1807?-85). Locomotive engineer; claimed to have run first locomotive built in U.S.

Am Eng & RR J (1884-5) 58:369-70

Pettit, Robert E. (? -1894). Engineer and railroad official.

RR Gaz (1895) 27:15

Pettit, William (1808-88). Mechanical engineer; worked on first locomotive of American construction.

Frank Inst J (1889) 127:294

Pharr, Henry N. (? -1897). Civil engineer; chief engineer of St. Francis levee district.

Eng Rec (1897) 36:465

Phelps, Edward Haight (1847-84). Railroad engineer.

RR Gaz (1884) 16:287

Phelps, George M. (1820-88). Inventor and instrument maker; improver of telegraph.

Elec World (1888) 11:268*
RR Gaz (1888) 20:342

Phelps, George M., Jr. (1843-95). Electrical engineer, important in development of the telegraph; president of weekly periodical Electrical Engineer.

AIEE Trans (1895) 12:664-5*
Elec Eng (1895) 19:355*
Eng Rec (1894-5) 31:363
Iron Age (1895) 55:824
RR Gaz (1895) 27:258
Sci Am (1895) 72:291

Philbrick, Edward Southwick (1827-89). Sanitary
    engineer; civil engineer, railroads and Hoosac
    Tunnel.

    Am Eng & RR J (1889) 63:145
    ASCE Proc (1897) 23:427-9
    ASCE Trans (1897) 38:454-5
    Assoc Eng Soc J (1889) 8:435-40
    Eng & Min J (1889) 47:170
    Eng Rec (1888-9) 19:157,166
    RR Gaz (1889) 21:133

Philippi, Adolph (? -1895). Civil engineer.

    Eng Rec (1894-5) 31:93

Phillips, John M. (1817-84). Manufacturer of steam
    engines, boilers, and machine tools; later pro-
    prietor of iron works.

    Am Mach (3/8/1884) 7:7
    Iron Age (2/21/1884) 33:13

Phillips, W. A. (? -1897). Mechanical engineer.

    Eng Rec (1897-8) 37:25

Phillips, William J. (1825-92). Inventor; improvements
    in telegraphic and electrical devices.

    Elec World (1892) 19:391

Phinney, Henry Ward Beecher (1854-88). Civil engineer,
    water works.

    ASCE Proc (1896) 22:696
    ASCE Trans (1896) 36:563

Phleger, Leonard (? -1881).   Inventor of coal
    burning locomotive boiler.

    RR Gaz (1881) 13:336

Picard, F. J. (? -1900). Civil engineer.

    Eng Rec (1900) 42:524

Pickering, D. N. (? -1886). Master mechanic of Boston
    & Worcester Railroad; general manager of Central
    Iowa Railroad.

    RR Gaz (1886) 18:65

Pickering, Thomas Richard (1831-95). Mechanical
    engineer.

    ASME Trans (1895) 16:1193-4

Pierce, Winslow S. (1819-88). Developer of canals and iron industry, Indiana.

Eng & Min J (1888) 46:89

Pierson, Frederick R. (1838?-98). Assayer and mining engineer.

Eng & Min J (1898) 65:138

Pike, William A. (1851-95). Mechanical and consulting engineer; Dean of School of Engineering, University of Minnesota.

ASCE Proc (1895) 21:212-3
ASME Trans (1896) 17:741
Eng Rec (1895) 32:363

Piolett, Victor E. (1812?-90). Construction superintendent for canals and railroads.

Eng Rec (1890) 22:194

Platt, George H. (1854-97). Railroad master mechanic.

ASME Trans (1897) 18:1104-5

Platt, George W. (1824-1900). Civil engineer.

Eng Rec (1900) 42:258

Platt, J. H. (? -1894). Oil refiner and paper manufacturer.

Eng & Min J (1894) 58:155

Platt, Joseph C. (1845-98). Consulting mechanical and hydraulic engineer; president of Mohawk and Hudson Manufacturing Company.

AIME Trans (1899) 29:xxxiii-iv
ASME Trans (1898) 19:982
Eng Rec (1898) 38:113

Poe, Orlando Metcalf (1832-95). Civil and military engineer, canal work in Great Lakes area.

AISA Bull (1895) 29:227
Eng & Min J (1895) 60:327
Eng Rec (1895) 32:327
Iron Age (1895) 56:747
RR Gaz (1895) 27:662
West Soc Eng J (1896) 1:128-30*

DAB 15:28-9
NYT (10/3/1895) 8:4

Pohle, Julius C. (1829-95). Mining engineer and chemist;
    inventor of Pohle air-lift pump.

    Am Mach (1895) 18:827
    Eng & Min J (1895) 60:351
    Eng Rec (1895) 32:345
    Iron Age (1895) 56:747
    RR Gaz (1895) 27:679

Poillon, Richard (? -1891). Shipbuilder.

    Am Eng & RR J (1891) 65:381

Pollock, Allan (1766?-1859). Inventor of stove.

    Sci Am (1859) 1:160

Pollock, Anthony (? -1898). Civil engineer and patent
    lawyer.

    Eng Rec (1898) 38:113

Pollock, William (1812?-92). Iron manufacturer and
    engineer.

    Eng & Min J (1892) 54:84

Pomeroy, Phinehas (1801?-86). Civil engineer, rail-
    roads in West.

    RR Gaz (1886) 18:170

Pond, Lucius M. (1826?-89). Manufacturer of machine
    tools.

    AISA Bull (1889) 23:157
    Am Mach (5/30/1899) 12:8
    Sci Am (1875) 33:273,311

Pope, Franklin Leonard (1840-95). Inventor of improve-
    ments in printing telegraph, electrical road
    signals; editor of Electrical Engineer.

    Am Eng & RR J (1895) 69:503
    AIEE Trans (1895) 12:670-83*
    Am Mach (1895) 18:845
    Elec Eng (1895) 20:413*
    Elec World (1895) 26:96*, 423,456
    Eng (1895) 30:103
    Eng & Min J (1895) 60:375
    Eng Rec (1895) 32:363
    Iron Age (1895) 56:302
    RR Gaz (1895) 27:694
    Sci Am (1895) 73:259

    Appleton's Cyc 5:67
    DAB 15:75-6

Pope, Willard Smith (1832-95). Civil engineer, bridges
    in Mississippi area.

    ASCE Proc (1896) 22:61-4
    ASCE Trans (1896) 36:565-8
    Eng Rec (1895) 32:363
    Iron Age (1895) 56:802
    RR Gaz (1895) 27:694
    West Soc Eng J (1896) 1:269-72*

Porter, Albert H. (? -1899). Civil engineer; worked
    on development of hydraulic power, Niagara Falls.

    Eng Rec (1899) 40:254

Porter, George A. (1845-92). Manufacturer of boilers.

    ASME Trans (1892) 13:672-3

Porter, Rufus (1792-1884). Inventor; founder of
    Scientific American.

    Sci Am (1884) 51:297

    Appleton's Cyc 5:79-80
    DAB 15:101-2

Post, Andrew J. (1834-96). Machinist and civil
    engineer, iron bridges and buildings.

    Eng Rec (1895-6) 33:255
    Iron Age (1896) 57:707
    RR Gaz (1896) 28:207

Post, Benjamin M. (1837?-95). Chief engineer of several
    steamship companies.

    Iron Age (1895) 55:1335

Post, James Clarence (1844-96). Military engineer;
    river and harbor improvements, waterworks in the
    Northwest.

    ASCE Proc (1896) 22:85-6
    ASCE Trans (1896) 36:569
    Eng & Min J (1896) 61:43
    Eng Rec (1895-6) 33:93
    RR Gaz (1896) 28:30

Post, Simon S. (1805-72). Civil engineer, railroads
    in N.Y. and Midwest; specialized in bridge
    construction; authored treatise on bridges.

    ASCE Proc (1893) 19:49-50

Potis, Salvator (1860-1900). Civil and mechanical consulting engineer.

ASME Trans (1900) 21:1167
Iron Age (4/26/1900) 65:21
RR Gaz (1900) 32:279

Potter, Charles (1824?-99). Inventor of Potter printing press.

Am Mach (1899) 22:1167

Potter, Thomas J. (1840-88). Railroad engineer, Union Pacific.

Ind World (3/15/1888) 30:5

Appleton's Cyc 5:91

Potts, John (1822?-98). Iron manufacturer; builder of machinery for handling anthracite.

Eng & Min J (1898) 66:674

Potts, Joseph D. (1829-93). Civil engineer, railroads; director of storage companies.

AISA Bull (1893) 27:357
ASCE Proc (1894) 20:49-51
Eng & Min J (1893) 56:598
Eng Rec (1893-4) 29:34
Iron Age (1893) 52:1030

Potts, Richard (1831-91). Civil engineer, public works, Chicago.

ASCE Proc (1891) 17:242

Power, Laurence (1838?-97). Manufacturer of woodworking machinery.

Iron Age (2/25/1897) 59:20

Powers, Thomas J. (1807?-88). Civil engineer, railroads and canals.

Am Eng & RR J (1889) 63:47

Pratt, F. G. (1860?-96). Inventor of electrical devices.

Elec Eng (1896) 22:604
Elec World (1896) 28:742

Pratt, Nathaniel W. (1852-96). Engineer and inventor; patented first successful dynamite gun.

Am Eng & RR J (1896) 70:61
AISA Bull (1896) 30:67
Am Manuf (1896) 58:407
ASME Trans (1896) 17:745-6
Elec Eng (1896) 21:301
Eng (1896) 31:79,90*
Eng Rec (1896-7) 35:273
Iron Age (1896) 57:707
Power (4/1896) 16:15
RR Gaz (1896) 28:207

Pratt, Thomas Willis (1812-75). Civil engineer, railroads; inventor of Pratt truss for bridges.

ASCE Proc (1873-5) 1:332-5

DAB 15:179

Prendergast, F. S. (? -1885). Consulting engineer, railroads.

RR Gaz (1885) 17:397

Prendergast, Francis E. (1841-97). Civil engineer, railroads; technical writer.

ASCE Trans (1898) 39:701-2
Eng Rec (1897-8) 37:47

Price, James Martin (1825-99). Deviser of railraod appliances.

Elec World (1899) 34:602

Price, John A. (1842?-92). Mechanical engineer and inventor; studied uses for coal wastes.

ASME Trans (1892) 13:683-4
Eng & Min J (1892) 54:132
Eng Rec (1892) 26:166
Iron Age (1892) 50:291

Prime, Frederick E. (1829-1900). Military engineer.

Eng Rec (1900) 42:163
RR Gaz (1900) 32:559

Prince, Frederick H. (1849-97). Manufacturer of elevators and hoisting machinery.

Iron Age (2/11/1897) 59:17

Prince, Philip M. (1848-94). Military engineer.

ASCE Proc (1894) 20:200-2
Eng Rec (1894) 30:302

Prosser, Thomas (1801?-70). Engineer; inventor and
manufacturer of boiler tubes.

ASCE Proc (1896) 22:611-2
ASCE Trans (1896) 36:564
RR Gaz (1870-1) 2:36

Prosser, Treat T. (1827-95). Mechanical engineer;
prominent in development of steam enginer, bolt
and screw industries.

Eng Rec (1895-6) 33:39
Iron Age (1895) 56:1329

Pullman, Albert Benton (1828-94). Manufacturer of
sleeping car.

Nat Car (1894) 25:10

Pullman, George Mortimer (1831-97). Inventor of
sleeping car; railroad car manufacturer.

Am Eng & RR J (1897) 71:387
AISA Bull (1894) 28:155
Am Manuf (1897) 61:594
RR Gaz (1897) 29:754
Sci Am (1897) 77:278

Appleton's Cyc 5:134*
DAB 15:263-4
NYT (10/20/1897) 4:1

Pusey, Joshua L. (1820-91). Shipbuilder and machinist.

Am Eng & RR J (1891) 65:139
Am Mach (3/5/1891) 14:8
Iron Age (1891) 47:292

Putnam, Charles A. (? -1899). Civil engineer.

Eng Rec (1899) 40:18

Putnam, Joseph W. (1836-93). Civil engineer, bridges
and railroads; designed and built apparatus for
handling and creosoting timber.

ASCE Proc (1894) 20:87

Queen, James W. (1812?-90). Manufacturer of delicate
instruments for scientists, surveyors and chemists.

Elec Eng (1890) 10:121
Elec World (1890) 16:62
Sci Am (1890) 63:101

Quick, Robert W. (? -1899). Professor of
physics and electrical engineering.

Elec World (1899) 34:511

Quillard, Claude Victor (? -1889). Civil engineer.

Eng & Min J (1889) 48:251

Radford, Benjamin F. (1827-94). Inventor of
machinery for processing sugar and leather.

ASME Trans (1895) 16:1189-90

Ramsay, Robert (1841?-99). Construction engineer.

Eng & Min J (1899) 68:284

Ramsey, Morris (? -1893). Mining engineer; inspector
and manager for Southwest Coal and Coke Company, Pa.

AISA Bull (1893) 27:5
Am Manuf (1893) 52:57
Iron Age (1/5/1893) 51:24

Rand, Addison Crittenden (1841-1900). Developer of
rock drills and air compressors.

Am Eng & RR J (1900) 74:126
Am Mach (1900) 23:259,281
Am Manuf (1900) 60:265
ASME Trans (1900) 21:1166
Eng & Min J (1900) 69:314*
Eng Rec (1900) 41:257
Iron Age (3/15/1900) 65:22
Mach (1899-1900) 6:254
RR Gaz (1900) 32:177

DAB 15:343-4

Rand, Jasper R. (1837-1900). Manufacturer of rock
drills.

Am Manuf (1900) 67:68*
Eng & Min J (1900) 70:76
RR Gaz (1900) 32:518

Randolph, James Lingan (1817-88). Consulting civil engineer, railroads; specialized in bridges.

Am Eng & RR J (1888) 62:525
ASCE Proc (1892) 18:217-9
Eng & Min J (1888) 46:243
RR Gaz (1888) 20:643

Randolph, Peyton (? -1891). Civil engineer.

Am Eng & RR J (1891) 65:283

Randolph, Richard (? -1893). Engineer, Baltimore Belt Line Tunnel.

Eng & Min J (1893) 55:132
Eng Rec (1892-3) 27:211

Rankin, James L. (1877?-1900). Assistant chief engineer, American Tin Plate Company.

Am Manuf (1900) 67:302

Rathbone, Henry (1813-91). President of iron and steel rolling mill; paper manufacturer.

Iron Age (1891) 48:596

Rauschenbach, August (1830?-99). City engineer, St. Louis.

Eng Rec (1899) 39:553

Rawle, Henry (1833-99). Civil engineer, railroads; later iron manufacturer and coal developer.

AISA Bull (1899) 33:213
Eng Rec (1899) 40:682

Appleton's Cyc 5:190

Read, John (? -1893). Civil engineer, railroads in Maine.

Eng Rec (1892-3) 27:506

Read, Nathan (1759-1849). Inventor of multi-tubular boiler, double acting steam engine and improved paddle wheel; iron manufacturer.

Sci Am (1870) 22:328

Appleton's Cyc 5:199-200
DAB 15:429-30

Reed, Charles F. (? -1898). Civil engineer and surveyor, railroads.

Eng Rec (1897-8) 37:271

Reed, Edward M. (1821-92). Civil engineer, railroads and bridges.

Am Eng & RR J (1892) 66:146
ASCE Proc (1892) 18:94-6
ASME Trans (1892) 13:676-7
Eng Rec (1891-2) 25:206
Nat Car (1892) 23:44

Reed, James H. (? -1891). Mechanical and marine engineer.

Am Mach (4/7/1891) 14:7

Reed, Samuel B. (1818-91). Civil engineer, Union Pacific and other railroads.

ASCE Proc (1892) 18:183-5
Eng Rec (1891-2) 25:70
RR Gaz (1892) 24:15

Reed, Thomas L. (? -1900). Inventor of flexible gaslight tubing and vulcanized rubber bulb.

Elec World (1900) 35:201

Reedy, James W. (? -1889). Elevator manufacturer.

Iron Age (1889) 44:610

Rees, James (1821-89). Pittsburgh industrialist; builder of first steel plate boats in U.S.; inventor of hot die press.

Am Manuf (9/20/1889) 45:11
Eng & Min J (1889) 48:228
Iron Age (1889) 44:449

DAB 15:464-5

Reeves, Samuel J. (1818-78). Civil engineer; made improvements in design and construction of wrought iron framing.

AISA Bull (1879) 13:324; (1878) 12:300
ASCE Proc (1879) 5:93-6
Eng News (1879) 6:6

Rehfus, George (1825?-1900). Manufacturer of light machinery.

Am Mach (1900) 23:985
Iron Age (10/11/1900) 66:28

Reid, James H. (? -1891). Marine engineer.

Eng (1891) 21:90

Reilly, Dennis (1836?-89). Contractor and builder of
bridges and roads; manufacturer of sheet iron.

AISA Bull (1889) 23:197
Iron Age (1889) 44:176-7

Remington, Philo (1816-89). Inventor; manufacturer
of firearms, sewing machines and typewriters.

Iron Age (1889) 43:555

Appleton's Cyc 5:219-20
DAB 15:498-9

Reno, James Hart (? -1881). Civil engineer, railroads
in West.

ASCE Proc (1897) 23:175
ASCE Trans (1897) 37:566

Renwick, Henry Brevoort (1817-95). Civil and
mechanical engineer, responsible for several
federal construction works; patent expert.

Am Mach (1895) 18:148
Eng & Min J (1895) 59:107
Eng Rec (1894-5) 31:165

Appleton's Cyc 5:222-3
DAB 15:505

Renwick, James (1792-1863). Mechanic and technical
writer.

Sci Am (1863) 8:58

DAB 15:506-7

Renwick, James (1818-95). Engineer and architect;
designed several celebrated buildings.

Eng & Min J (1895) 59:611

DAB 15:507-9
NYT (6/25/1895) 9:5

Reynolds, George F. (1864-91). Mining engineer.

ASME Trans (1892) 13:672

Rhett, Thomas S. (? -1893). Engineer, employed in
building of Mexican National Railroad.

Eng Rec (1893-4) 29:68

Rhodes, Benjamin (1849-94). Civil engineer, hydraulic work and bridge construction, Niagara Falls, New York.

ASCE Proc (1894) 20:163-4
Eng Rec (1894) 30:182

Rice, Edward Curtis (1829-98). Civil engineer, railroads.

ASCE Trans (1898) 39:703

Rice, Frederick B. (1851-89). Mechanical engineer; inventor of oil engine and Rice automatic compound engine.

ASME Trans (1890) 11:1152

Richards, George (? -1889). Mining engineer.

Eng & Min J (1889) 47:461

Richards, John (? -1880). Mechanical engineer and manufacturer; introduced American woodworking machinery to Europe.

Am Mach (3/6/1880) 3:9
RR Gaz (1880) 12:133,248

Richardson, F. W. (1852?-86). Patentee and manufacturer of Richardson-Allen balance slide valve and muffler for safety valve.

Am Manuf (2/5/1886) 38:11
RR Gaz (1886) 18:82

Richardson, Frank E. (1860?-92). Civil engineer, employed in iron works.

Eng Rec (1891-2) 25:274

Richardson, George (? -1898). Inventor and waterworks manager.

Eng Rec (1898-9) 39:61

Richardson, George W. (1828-92). Railroad engineer and inventor of pop safety valve.

Am Mach (9/15/1892) 15:8

Richardson, John Owen (1851-97). Mechanical engineer.

Elec Eng (1897) 23:17

Richardson, Seneca M. (1830?-99). Manufacturer of woodworking machinery, Worcester, Mass.

Am Mach (1899) 22:826
Iron Age (8/24/1899) 64:17

Richmond, Dennison (1842?-88). Division engineer, state canals of New York.

Eng Rec (1888) 18:227

Rickard, Michael (? -1896). Locomotive engineer, State Railroad Commission of New York.

RR Gaz (1896) 28:888

Ricker, Robert E. (? -1894). Consulting engineer, railroads.

Am Eng & RR J (1894) 68:335
Eng Rec (1893-4) 29:408
Nat Car (1894) 25:92
RR Gaz (1894) 26:379

Ridall, John E. (1850?-98). Developer of electric lighting.

Elec Eng (1898) 25:195

Riddle, Hugh (1822-92). Civil engineer, railroads.

Am Eng & RR J (1892) 66:432
Eng Rec (1892) 26:180
RR Gaz (1892) 24:624

Rider, Alexander K. (1821-93). Mechanic; inventor of Rider hot-air pumping and cut-off engines.

Am Mach (10/5/1893) 16:4-5
Eng Rec (1893) 28:246
Iron Age (1893) 52:521

Ridgway, Joseph T. (1838-96). Mechanical engineer.

ASME Trans (1897) 18:1092

Riehle, Henry B. (? -1890). Manufacturer of scales and testing machines.

Am Eng & RR J (1890) 64:281

Riotte, Eugene N. (1843-91). Mining engineer and metallurgist.

Eng & Min J (1891) 51:583*

Rittenhouse, Charles Tomlinson (1871-1900).
Experimenter in electricity and X-ray photography;
editor of Electrical World.

Elec World (1900) 35:384
Eng Rec (1900) 41:233

Roach, John (1813 or 1815-87). Father of iron ship-
building in America.

Am Eng & RR J (1887) 61:53
AISA Bull (1887) 21:12,19
Am Manuf (1/28/1887) 40:6
Elec Eng (1887) 6:43
Eng & Min J (1887) 43:46
Ind World (1/13/1887) 28:1*,5-6
Iron Age (1/13/1887) 39:15,26
RR Gaz (1887) 19:30
Sanit Eng (1886-7) 15:166
Sci Am (1882) 47:19-20*; (1887) 56:48

Appleton's Cyc 6:268-9*
DAB 15:639-40
NYT (1/11/1887) 3:1

Robbins, Thomas C. (1830?-96). Developed electrical
motive power for railroads.

Elec World (1897) 29:80

Roberts, E. A. L. (? -1881). Engineer and inventor
of torpedo for oil wells.

RR Gaz (1881) 13:188
Sci Am (1881) 44:241

Roberts, George Brodee (1833-97). Civil engineer,
railroads; President of Pennsylvania Railroad.

Am Eng & RR J (1897) 71:100
AISA Bull (1897) 31:37
Am Mach (1897) 20:122
Am Manuf (1897) 60:191
Eng & Min J (1897) 63:137*
Eng Rec (1896-7) 35:201
Iron Age (2/4/1897) 59:21
Mach (1896-7) 3:226
RR Gaz (1897) 29:98-9

DAB 16:6-7

Roberts, Howard N. (1861-91). Railroad engineer.

Eng Rec (1891) 24:380

Roberts, Percival (1830-98). Proprietor of Pencoyd
    Iron Works, first iron bridge building firm in
    the U.S.

    Am Mach (1898) 21:264
    Eng & Min J (1898) 65:408
    Iron Age (4/7/1898) 61:18*
    RR Gaz (1898) 30:264

Roberts, Solomon White (1811-82). Civil engineer,
    canals and railroads in Pa.; operated first
    successful anthracite furnace in U.S.

    RR Gaz (1882) 14:200

    Appleton's Cyc 5:276
    DAB 16:15-6

Roberts, W. B. (1838-89). Inventor of torpedoes for
    blasting oil wells.

    Am Manuf (8/2/1889) 45:11

Roberts, Willard B. (1848-1900). Mechanical engineer.

    ASME Trans (1900) 21:1166
    RR Gaz (1900) 32:279

Roberts, William Milnor (1810-81). Civil engineer,
    canals and railways; chief engineer, enlargement
    of Erie Canal; a president of the ASCE.

    Am Eng & RR J (1881) 54:868
    Am Manuf (7/29/1881) 29:10
    ASCE Proc (1896) 22:683-9
    ASCE Trans (1896) 36:531-7
    Eng News (1881) 8:291,301
    Iron Age (7/28/1881) 28:15
    RR Gaz (1881) 13:416-7
    Sci Am (1881) 45:115
    Van Nostrand's (1881) 25:255

    Appleton's Cyc 5:276-7
    DAB 16:18-9
    NYT (7/31/1881) 2:7

Robertson, John (1809-96). Mechanical engineer and
    inventor of machinery; proprietor of Tubal Cain
    Iron Works.

    Eng Rec (1896) 34:81
    Iron Age (1896) 58:68

Robinson, A. P. (1822-98). Railroad and civil engineer; involved in survey of Erie Railroad and in construction of railways in Southwest.

Eng Rec (1898) 38:465
Iron Age (10/27/1898) 62:21

Robinson, J. R. (1823-91). Steam engineer.

Eng & Min J (1890) 50:265

Robinson, John M. (1835-93). Civil engineer, railroads and steamships; coal operator.

AISA Bull (1893) 27:59
Eng & Min J (1893) 55:156
Eng Rec (1892-3) 27:231

Robinson, Lester L. (? -1892). Railway engineer; developed hydraulic mining.

Eng Rec (1891-2) 25:374

Robinson, Moncure (1802-91). Civil engineer; prominent in development of railroads.

Am Eng & RR J (1891) 65:574
ASCE Proc (1892) 18:84-90
Eng & Min J (1891) 52:502
Eng News (1889) 21:335-6*; (1891) 26:463
Eng Rec (1891) 24:380
RR Gaz (1891) 23:797-8

DAB 16:48-9

Roche, John (? -1894). Civil engineer, employed on Wabash & Erie Canal.

Eng Rec (1894) 30:354

Roderick, Matthew (? -1896). Civil engineer, Seattle.

Eng Rec (1896) 34:399

Rodman, Thomas J. (1815-71). Soldier and inventor of processes for casting guns and projectiles.

Sci Am (1871) 25:103

Appleton's Cyc 5:298-9
DAB 15:80-1
NYT (6/8/1871) 5:5

Roebling, John Augustus (1806-69). Civil engineer
and wire rope manufacturer; builder of suspension
bridges; designer and planner of the Brooklyn
Bridge.

Eng News (1883) 10:246*
Frank Inst J (1867) 84:410-3
Sci Am (1869) 21:89
Van Nostrand's (1869) 1:862-3

Appleton's Cyc 5:303
DAB 16:86-9
NYT (7/23/1869) 4:6

Rogers, Albert Brainerd (1829-89). Engineer, railroads
in the Rocky Mountains.

ASCE Proc (1890) 16:98-9

Rogers, Fairman (1833-1900). Civil engineer and
professor, University of Pennsylvania.

Eng & Min J (1900) 70:256
Eng Rec (1900) 42:186

Appleton's Cyc 5:305

Rogers, Henry J. (1811?-79). Inventor; devised signal
codes and telegraphic instruments.

Sci Am (1879) 41:161

Appleton's Cyc 5:305-6
DAB 16:96-7

Rogers, John Benjamin (1832-74). Mechanical engineer
and inventor, India; railroad engineer, St. Louis.

ASCE Proc (1873-5) 1:168-9

Rogers, Thomas (1792-1856). Designer and builder
of locomotives; manufacturer of cotton handling
machinery.

Am Eng & RR J (1856) 29:280

DAB 16:112-3

Rogers, William Barton (1804-82). Geologist; one of
founders of the Massachusetts Institute of Technology.

Sci Am (1882) 46:393

Appleton's Cyc 5:306-7*
DAB 16:115

Roosevelt, Hilbourne Lewis (1849-86). Organ builder, electrician and inventor; applied electricity to organ movements.

Sci Am (1887) 56:49

DAB 16:133

Root, Francis M. (1824?-89). Inventor of Root rotary blower.

Eng & Min J (1889) 48:390
Iron Age (1889) 44:685

Root, John B. (1830-87). Mechanical engineer and inventor of steam boilers.

Am Mach (1/1/1887) 10:8
ASME Trans (1887) 8:726-7
Sanit Eng (1886-7) 15:70

Root, Lewis F. (1828?-93). Civil engineer, Massachusetts.

RR Gaz (1893) 25:270

Root, Porteous B. (? -1882). Civil engineer, railroads.

RR Gaz (1882) 14:121

Root, William J. (1844-92). Engineer, hydraulic works.

ASME Trans (1893) 14:1443

Roots, Francis M. (1824-90). Inventor of rotary blower, gas exhauster and steam pump.

Sci Am (1890) 62:6

Roots, P. H. (1813-79). Mechanical engineer; developed and perfected rotary blower.

Sci Am (1879) 40:385

Roper, H. S. (1823?-96). Machinist and inventor of steam-propelled vehicles.

Am Mach (1896) 19:589

Rose, George M. (? -1893). Engineer and metallurgist.

RR Gaz (1893) 25:860

Rose, Joshua (1837?-98). Mechanical engineer and prolific technical writer.

Am Mach (1898) 21:888
ASME Trans (1899) 20:1001
Sci Am (1898) 79:371

Rosecrans, William Starke (1819-98). Soldier and
  military engineer; mining engineer.

  Am Eng & RR J (1898) 72:131
  AIME Trans (1899) 29:xxxiv
  RR Gaz (1898) 30:207

  Appleton's Cyc 5:323-4*
  DAB 16:163-4
  NYT (3/12/1898) 7:3

Roser, Thomas Lafayette (1836-98). Civil engineer,
  Northern Pacific and Canadian Pacific Railroads.

  RR Gaz (1898) 30:444

Rotham, A. L. (1838-85). Civil engineer, steel works.

  Sanit Eng (1884-5) 11:378

Rothman, A. L. (1838-85). Civil engineer, involved
  with introduction of Bessemer process in U.S.

  Am Manuf (3/27/1885) 36:14
  Iron Age (3/19/1885) 35:19

Rowe, Robert Delos (1863-99). Consulting and civil
  engineer, railroads.

  ASCE Proc (1899) 25:745
  ASCE Trans (1899) 42:571

Rowland, Frank L. (? -1888). Engineer, New York and
  Brooklyn Bridge.

  Eng Rec (1888) 18:266

Royal, George (1837-97). Machinist and master
  mechanic.

  Am Eng & RR J (1897) 71:99

Royce, Frederick W. (1839?-1900). Inventor and
  electrician.

  Elec World (1900) 36:828

Rubel, M. (1834?-86). Manufacturer of cutlery and
  cutlery-producing machinery.

  Ind World (3/18/1886) 26:6

Rudd, C. H. (? -1894). Inventor of electrical devices.

  Elec Eng (1894) 18:135

Rudloff, Henry Frederick (1846-95). Railroad engineer; worked primarily in Latin America.

ASCE Proc (1896) 22:102
ASCE Trans (1896) 36:570

Ruggles, S. P. (? -1880). Inventor of Ruggles printing press, an early machine press.

Sci Am (1880) 43:7

NYT (5/31/1880) 2:2

Ruggles, Thomas Colden (1814?-93). Civil engineer, Harlem Railroad.

Eng & Min J (1893) 56:350
Eng Rec (1893) 28:278

Rushforth, William Henry (1844?-92). Inventor of railroad appliances.

Eng Rec (1892) 26:194

Russell, Howland (1853?-92). Mining engineer.

Eng Rec (1891-2) 25:342

Russell, William D. (? -1876). President of Baxter Steam Engine Company, Newark, N.J.

Sci Am (1876) 35:22

Russell, William H. (1818?-94). Consulting engineer, Boston & Albany Railroad.

Eng Rec (1894-5) 31:40
RR Gaz (1894) 26:361

Rust, James F. Putnam (1819?-99). Naval engineer.

Am Mach (1899) 22:311

Rust, John R. (1828?-99). Railway engineer.

Eng Rec (1899) 40:370

Rutherford, William H. (1828-98). Chief Engineer, U.S. Navy.

Am Mach (1898) 21:192

Rutter, James (? -1895). Civil engineer and railroad contractor.

RR Gaz (1895) 27:304

Rutter, Thomas (? -1895). Civil engineer, railroads and tunnels.

Eng & Min J (1895) 59:443
Eng Rec (1894-5) 31:417

Sabbaton, Frederic A. (? -1894). Engineer, specialized in gas works construction.

Eng Rec (1893-4) 29:328

Sabbaton, Paul A. (1788?-1869). Gas engineer.

Sci Am (1869) 21:345

Sague, Samuel L. (1841-1900). Inventor of appliances for steel manufacturing.

Iron Age (3/8/1900) 65:19

St. John, H. Roswell (1844?-1900). Inventor of typesetting machinery.

Am Mach (1900) 23:753
Iron Age (8/2/1900) 66:20

St. John, Isaac Munroe (1827-80). Civil engineer; railroads in South and Midwest; city engineer of Louisville, Kentucky.

ASCE Proc (1896) 22:694-5
ASCE Trans (1896) 36:571

Appleton's Cyc 5:371
DAB 16:302-3

Sage, Russell (? -1892). Tunnel and railway engineer.

Eng Rec (1891-2) 25:154

Samuel, Edward (1845-96). Manufacturer of iron and steel; locomotive builder.

AISA Bull (1896) 30:76
Eng & Min J (1896) 61:355
Ind World (4/9/1896) 46:15
Iron Age (1896) 57:820

Sandford, C. O. (1811-83). Civil engineer and railroad builder, South.

RR Gaz (1883) 15:832
Sci Am (1883) 49:353

NYT (11/30/1883) 4:7

Sands, Thomas (1833-1900). Inventor and manufacturer.

Iron Age (5/3/1900) 65:27

Sargent, Edward (1826-1900). Railway engineer.

Eng Rec (1900) 42:552

Sawyer, Charles L. (? -1898). Mechanical engineer.

Am Mach (1898) 21:848

Sawyer, Joseph Bartlett (? -1897). Civil engineer,
New Hampshire.

Eng Rec (1897) 36:3

Sawyer, William E. (? -1883). Electrical engineer.

Elec World (1883) 1:307

Saxton, Joseph (1799-1873). Inventor of clocks,
balances, coal-burning stove and other devices.

Sci Am (1869) 21:394

Appleton's Cyc 5:409
DAB 16:400
NYT (10/29/1873) 6:7

Schaufuss, Erich C. (1856-89). Mining engineer.

AIME Trans (1888-9) 17:419-20

Scherzer, William (1858-93). Consulting and contracting
bridge engineer.

ASCE Proc (1894) 20:58-9
Assoc Eng Soc J (1894) 13:227-8
Eng Rec (1893) 28:151

Schoen, William H. (? -1896). Manufacturer of
car and locomotive springs.

RR Gaz (1896) 28:367

Schoenleber, John J. (1866-98). Electrical engineer.

ASME Trans (1898) 19:981-2
Elec Eng (1898) 26:161

Schofield, William (1831-1900). Manufacturer of
textile machinery.

Am Manuf (1900) 23:360
Iron Age (4/19/1900) 65:32

Schott, Charles Henry (1847-78). Surveyor for
  railroads and civic works.
    ASCE Proc (1878) 4:128-9

Schubert, Julius (? -1898). Introduced improvements
  in machinery for asphalt refining and paving.
    Eng Rec (1898) 38:311

Schultz, William H. (1805?-99). Engineer and inventor.
    Am Mach (1899) 22:37

Schulz, Carl J. (? -1897). Civil engineer, waterworks.
    Eng Rec (1897) 36:179

Schulze-Berge, Franz (1856-94). Experimenter in
  electricity; inventor of rotary mercury vacuum pump.
    AIEE Trans (1894) 11:873; (1895) 12:665

Schuyler, Howard (1844-83). Civil engineer, railroads.
    ASCE Trans (1896) 36:572-4
    Assoc Eng Soc J (1883) 2:297-9
    RR Gaz (1883) 15:816

Schuyler, John (1829?-95). Civil engineer and bridge
  builder.
    Eng Rec (1895) 32:219
    Iron Age (1895) 56:392

Schwanecke, H. O. (? -1894). Railroad engineer.
    Eng Rec (1893-4) 29:232
    RR Gaz (1894) 26:183

Scollay, John A. (? -1896). Heating and ventilating
  engineer.
    Eng Rec (1896) 33:327

Scott, Chalmers (? -1898). Civil engineer, San Diego.
    Eng Rec (1898-9) 39:3

Scott, Harry C. (1867?-98). Civil engineer, bridges
  in Pa.
    Eng Rec (1897-8) 37:161
    RR Gaz (1898) 30:49

Scott, John (1819?-90). Engineer; a reputed inventor
  of the locomotive cab.
    Nat Car (1890) 21:186

Scott, Walter (? -1896). Civil engineer, railroads; builder of first pneumatic engine to transport natural gas.

Eng Rec (1896-7) 35:91
Iron Age (1896) 58:1318

Scott, William Ulysses (1870-98). Railroad engineer.

ASCE Proc (1899) 25:746
ASCE Trans (1899) 42:572

Scowden, Random C. (? -1893). Engineer, waterworks at Poughkeepsie, N.Y.

Eng & Min J (1893) 56:622
Eng Rec (1893-4) 29:34

Scranton, William H. (1840-89). Civil and mining engineer.

AIME Trans (1889) 18:213
ASME Trans (1889) 10:834
Eng & Min J (1889) 48:11*,316-7

Seabrook, Thomas (1817?-97). Civil engineer, railroads in Pa.

Am Eng & RR J (1897) 71:133
AISA Bull (1897) 31:61
Eng Rec (1896-7) 35:289
RR Gaz (1897) 29:176

Seaman, William (? -1892). Civil engineer, viaducts.

RR Gaz (1892) 24:182

Sedgley, James (1824?-92). Railroad master mechanic.

Am Eng & RR J (1892) 66:146
Nat Car (1892) 23:31

Seeley, Thomas Jennings (1848-83). Civil engineer, railroads.

ASCE Proc (1896) 22:691-2
ASCE Trans (1896) 36:574-5
RR Gaz (1883) 15:714

Seiler, D. W. (1835?-92). Civil engineer.

Eng Rec (1891-2) 25:222

Selby, W. H. (1832?-94). Superintendent of motive power.

Nat Car (1895) 26:10

Selden, George (1827-93). Inventor of sawmill machinery for oil works and of details of boilers and engines.

ASME Trans (1894) 15:1191
Iron Age (1893) 52:940

Selden, L. Hart (? -1891). Consulting engineer.

Am Eng & RR J (1892) 66:48
Eng Rec (1891) 25:22

Selden, Samuel H. (1836?-91). Civil engineer.

Eng Rec (1891-2) 25:2

Sellers, George Escol (1808-99). Manufacturer of locomotives and machinery.

Am Mach (1899) 22:250-1*

Sellers, George H. (1828-97). Inventor; civil engineer, bridges and hydraulic works.

AISA Bull (1897) 31:148,153
Eng Rec (1897) 36:25

Sennett, George Burritt (1840-1900). Manufacturer of oil-well machinery.

AISA Bull (1900) 34:67
Am Mach (1900) 23:305
Am Manuf (1900) 66:250

DAB 16:585-6

Serrell, James E. (1821?-92). Surveyor, N.Y. City.

Eng Rec (1892) 26:38

Sewell, George (? -1895). Chief engineer, U.S. Navy.

Eng (1895) 29:75

Seymour, H. C. (? -1853). Engineer, Erie Railroad.

Am Eng & RR J (1853) 26:488

Seymour, Louis Irving (1860?-1900). Mining engineer, primarily in Venezuela and South Africa.

Eng & Min J (1900) 69:746; 70-64*
Eng Rec (1900) 41:601
Iron Age (6/21/1900) 65:26
RR Gaz (1900) 32:433

Seymour, Mark Tucker (1820-85). Civil engineer and contractor, railroad bridges.

ASCE Proc (1897) 23:178
ASCE Trans (1897) 37:578

NYT (6/1/1885) 5:2

Seymour, Silas (1817-90). Civil engineer, canals and railroads.

Am Eng & RR J (1890) 64:379
Eng & Min J (1890) 50:80
Eng Rec (1890) 22:98
Iron Age (1890) 46:181
RR Gaz (1890) 22:532

Shallenberger, Oliver Blackburn (1860-98). Electrical engineer and inventor of electric meter.

AIEE Trans (1898) 15:744-53
AISA Bull (1898) 32:28
Elec Eng (1898) 25:116
Elec World (1898) 31:174*,206; (1899) 33:424
Eng Rec (1897-8) 37:183
Iron Age (2/3/1898) 61:26
RR Gaz (1898) 30:69

Shanahan, James (? -1897). Railroad contractor; superintendent of public works, New York State.

Eng Rec (1896-7) 35:333

Shanly, William (1819-1900). Civil engineer, railroads, canals, and tunnels.

Eng & Min J (1900) 69:194*

Sharp, Joel (1820-98). Machinist and machine shop owner; manufacturer of wire nails.

Am Mach (1898) 21:625
ASME Trans (1898) 19:981
Cassier's (1898) 14:544*
Eng (10/15/1898) 35:5

Sharp, Thomas M (1807?-97). Civil engineer.

Eng Rec (1897) 36:157

Sharpe, Lucian (1830-99). Head of Brown and Sharpe Manufacturing Company.

Am Mach (1899) 22:1026; 1069-70*
Mach (1899-1900) 6:92

Shaw, Albert M. (? -1889). Civil engineer, roads in New England.

Eng Rec (1888-9) 19:138

Shaw, Alton J. (1858-95). Mechanical engineer; inventor of electric crane.

Am Mach (1895) 18:529
ASME Trans (1895) 16:1190
Elec Eng (1895) 20:24

Shaw, Frank (1841-89). Electrician and inventor.

Elec Eng (1889) 8:192

Shaw, Joshua (1776-1860). Inventor.

Sci Am (1869) 21:90

Sheafer, Peter Wenrick (1819-91). Mining engineer.

Iron Age (1891) 47:640

Appleton's Cyc 5:489

Sheble, Franklin (1866-99). Electrical engineer.

Elec World (1899) 33:566*

Shedd, Charles Elmer (1858-92). Civil engineer and surveyor.

Eng Rec (1891-2) 25:428

Shelbourne, Sidney F. (? -1887). Civil and electrical engineer; inventor of wire insulators.

AIEE Trans (1887-8) 5:65-8
Elec Eng (1887) 6:501
Elec World (1887) 10:271

Sheldon, Frank W. (? -1897). Civil engineer.

Eng Rec (1897) 36:443

Shelly, Frederick S. (? -1897). Civil engineer and surveyor.

Eng Rec (1897) 36:157

Shelton, George Wellington (? -1890). Inventor of machine for bending wire.

Iron Age (1890) 46:534

Shepard, Charles R. (? -1899). Engineer and one of
   promoters of Erie Canal.

   Eng Rec (1899) 40:419

Sheriffs, James (1822-87). Operator of foundry;
   manufacturer of castings, compound engines and
   marine work.

   ASME Trans (1887) 8:729

Sherman, Joseph (? -1898). Civil engineer.

   Eng Rec (1898) 39:105

Sherman, William F. (? -1890). Inventor of Sherman
   storage battery system for street railway propulsion.

   Elec World (1890) 16:462

Sherwin, T. W. (1827?-87). Civil engineer.

   Am Eng & RR J (1881) 61:495

Sherwin, W. F. (? -1889). Surveyor; water commissioner
   of Erie, Pa.

   Sanit Eng (1887) 16:494

Shields, Charles O. (? -1896). Mining engineer.

   Eng & Min J (1896) 62:611

Shinn, William Powell (1834-92). Civil engineer and
   railroad manager; president of ASCE.

   Am Eng & RR J (1892) 66:289
   AIME Trans (1892-3) 21:394-400
   AISA Bull (1892) 26:129
   Am Manuf (1892) 50:850
   ASCE Proc (1892) 18:123-9
   Eng & Min J (1892) 53:495*
   Eng News (1890) 24:545*
   Eng Rec (1891-2) 25:374
   Iron Age (1892) 49:928-9*
   RR Gaz (1892) 24:357

Shippen, W. W. (1827?-85). Civil engineer.

   RR Gaz (1885) 17:572

Shoenberger, John H. (1809-89). Pioneer iron
   manufacturer, Pittsburgh.

   AISA Bull (1889) 23:313
   Am Manuf (11/15/1889) 45:11
   Iron Age (1889) 44:807

Shorey, George (? -1900). Electrical engineer.
  Elec World (1900) 36:397

Short, Samuel (? -1896). Engineer, Brooklyn.
  Eng Rec (1896) 34:307

Shortt, L. H. (? -1895). Civil engineer, San Francisco.
  Eng Rec (1895) 32:165

Shotwell, Robert M. (? -1894). Civil engineer.
  Eng Rec (1894-5) 31:93

Shreve, Samuel Henry (1829-84). Civil engineer, worked
  on rapid transit in New York and Brooklyn; author of
  treatises on bridges.
  ASCE Proc (1896) 22:86-7
  ASCE Trans (1896) 36:576-7
  Iron Age (12/4/1884) 34:23
  RR Gaz (1884) 16:874
  Sanit Eng (1884-5) 11:19

  NYT (11/29/1884) 2:6

Shriver, Harry Courtney (? -1896). Mechanical and
  electrical engineer, connected with development
  of rapid transit in Baltimore.
  Eng Rec (1895-6) 33:219

Shriver, Joseph (1806?-86). Civil engineer.
  RR Gaz (1886) 18:29

Sibley, Frederick (1822-89). Mechanical engineer.
  Eng & Min J (1889) 47:329
  Iron Age (1889) 43:555

Sibley, Hiram (1807-88). Machinist; one of organizers
  of Western Union Telegraph Co.; founder of Sibley
  College of Mechanical Arts, Cornell.
  Am Eng & RR J (1888) 62:381
  Am Mach (8/4/1888) 11:3
  Elec Eng (1888) 7:480
  Eng & Min J (1888) 46:28
  RR Gaz (1888) 20:480
  Sci Am (1888) 59:33*

  DAB 17:145-6
  NYT (7/13/1888) 5:1

Sickels, Frederick Ellsworth (1819-95). Engineer;
inventor of cut-off valve mechanism for engines
and steam-steering apparatus.

Am Mach (1895) 18:228,261-4*
ASCE Proc (1896) 22:130-4
ASCE Trans (1896) 36:577-82
Eng (1895) 29:86*
Eng Rec (1894-5) 31:273
Sci Am (1869) 21:409; (1895) 72:195

DAB 17:149-50

Sickels, Theophilus E. (1822-85). Civil engineer,
railroads and bridges.

ASCE Proc (1885) 11:130-1
Assoc Eng Soc J (1886-7) 6:232-5
RR Gaz (1885) 17:94
Sci Am (1885) 52:112

Sidell, William Henry (1810-73). Soldier and civil
engineer, bridges and railroads; chief engineer,
Panama Railroad.

ASCE Proc (1873-5) 1:41-2

Appleton's Cyc 5:524
DAB 17:151-2

Silliman, Justus Mitchell (1842-96). Mining
engineer and instructor at Lafayette College.

Eng & Min J (1896) 61:379
Eng Rec (1896) 33:345
Iron Age (1896) 57:977
RR Gaz (1896) 28:296

Silver, Thomas (1813-88). Inventor of governor for
marine steam engines; civil engineer.

Eng & Min J (1888) 45:292
RR Gaz (1888) 20:262

Appleton's Cyc 6:530
DAB 17:165-6

Simonds, George Frederick (1843-94). Inventor of
rolling process for steel.

Am Mach (11/29/1894) 17:11
ASME Trans (1895) 16:1189
Eng Rec (1894) 30:388

Simpson, James E. (1813-97). Shipbuilder; inventor
   and originator of timber graving dry docks.

   Eng Rec (1897) 36:465
   Sci Am (1897) 77:323

Sims, Alfred (1826?-95). Engineer and superintendent
   of railroads.

   Eng & Min J (1895) 59:491
   RR Gaz (1895) 27:287

Sinclair, James (? -1881). Inventor.

   Sci Am (1881) 44:241

Singer, Isaac Merritt (1811-75). Inventor and
   manufacturer of first practical domestic sewing
   machine, and of wood and metal carving machine.

   Sci Am (1875) 33:101

   Appleton's Cyc 5:542
   DAB 17:188-9
   NYT (7/25/1875) 6:6

Sites, Wilmon W. C. (1849-85). Civil engineer,
   railroads and canals; chief engineer, Jersey City
   Public Works Department.

   ASCE Trans (1896) 36:582-4
   RR Gaz (1885) 17:653

Sitgreaves, Lorenzo (? -1888). Member, Corps of
   Engineers.

   Eng Rec (1888) 18:10

Skeel, Theron (? -1878). Civil engineer.

   Eng & Min J (1878) 25:287

Skidmore, William B. (1822?-99). Superintending
   engineer, Brooklyn and New York Ferry Company.

   Am Mach (1899) 22:341

Skinner, Halcyon (1824-1900). Inventor of looms for
   carpets.

   Am Mach (1900) 23:1171

   DAB 17:198-9

Slater, Samuel (1768-1835). Machinist and early
   cotton manufacturer.

   Sci Am (1890) 63:233*

   DAB 17:205-6

Slater, Thomas (? -1895).   Civil engineer and surveyor.
   Eng Rec (1894-5) 31:183

Slattery, Marmaduke M. M. (? -1892).   Inventor of
   electrical alternating system and apparatus.
   Elec Eng (1892) 14:607

Slaven, Moses Albert (1846?-86).   Engineer and
   contractor.
   Sci Am (1886) 55:257

Slaymaker, Samuel C. (1827?-94).   Civil engineer,
   railroads in Lancaster, Pa.
   Eng & Min J (1894) 57:132
   Eng Rec (1893-4) 29:168
   RR Gaz (1894) 26:112

Slayton, Phineas (1820?-72).   Inventor.
   Sci Am (1872) 27:392

Sloss, James Withers (1820?-90).   Pioneer iron and
   steel manufacturer, Birmingham, Alabama.
   Am Manuf (5/16/1890) 46:13
   Iron Age (1890) 45:776,955-6*
   RR Gaz (1890) 22:347
   DAB 17:219

Small, David E. (1824-83).   Car manufacturer.
   RR Gaz (1883) 15:222

Small, Henry S. (? -1892).   Locomotive engineer.
   Nat Car (1892) 23:158

Smart, Charles E. (? -1896).   Master mechanic of
   Michigan Central Railroad.
   RR Gaz (1896) 28:242

Smedley, Samuel Lightfoot (1832-94).   Chief engineer
   and surveyor, Philadelphia.
   ASCE Proc (1895) 21:88-90
   Eng Rec (1894) 30:134

Smeed, Eben Cedron (1831-92).   Chief engineer, Union
   Pacific railroad system.
   Am Eng & RR J (1892) 66:479-80
   RR Gaz (1892) 24:676

Smith, Addison (? -1893).   Inventor.

   Eng Rec (1893-4) 29:18

Smith, Alba F. (? -1879).   Engineer, Hudson River
   Railroad.

   RR Gaz (1879) 11:406

Smith, Alfred Henry (1867-96).   Steam engineer and
   marine architect.

   ASME Trans (1896) 17:747

Smith, Charles Augustus (1846-94).   Civil and
   consulting engineer, railways, waterworks, etc.;
   professor, Washington University.

   ASCE Proc (1885) 11:122-3
   Assoc Eng Soc J (1884) 4:168-71
   RR Gaz (1884) 16:136
   Sci Am (1884) 50:153

Smith, Charles C. (1830-89).   Railroad engineer and
   designer of railroad bridges; chief engineer, St.
   Paul, Minneapolis and Manitoba Railroad.

   ASCE Proc (1889) 15:142-3

Smith, Charles D. (1855-90).   Mechanical engineer.

   ASME Trans (1890) 11:1154

Smith, Charles M. (? -1896).   Developed "electro-
   pneumatic" system of lighting and extinguishing
   street gas lamps.

   Eng Rec (1896) 34:177

Smith, Charles Shaler (1836-86).   Civil engineer,
   specialized in bridges; early ASCE director.

   Am Eng & RR J (1887) 61:6
   ASCE Proc (1887) 13:105-10
   RR Gaz (1886) 18:914
   Sanit Eng (1886-7) 15:94

   Appleton's Cyc 5:559
   DAB 17:251-2

Smith, Charles Vandervoort (1837-84).   Inventor of
   improvements in machinery for manufacture of
   illuminating gas.

   ASCE Proc (1885) 11:126-7

Smith, David Lowber (1846-93). Civil and mining
  engineer, involved with public works in N.Y. City
  and Brooklyn.

  ASCE Proc (1894) 20:57-8
  Eng Rec (1893) 28:310

Smith, David Reese (? -1897). Civil engineer and
  railroad developer.

  Eng Rec (1896-7) 35:157

Smith, Edward (1829-95). Engineer, railroads in Pa.

  Nat Car (1895) 26:140

Smith, Erastus Washington (? -1882). Mechanical
  and consulting engineer; designer of marine engines
  and of waterworks for New Orleans and Cleveland.

  Am Mach (7/8/1882) 5:1
  Eng (1882) 4:7
  RR Gaz (1882) 14:365
  Sci Am (1882) 47:3*

  NYT (6/13/1882) 5:1

Smith, Frederick Henry (1839-98). Geologist; civil
  engineer, bridges.

  ASCE Proc (1899) 25:331-2
  ASCE Trans (1899) 41:643-4

Smith, Frederick H. (? -1898). Civil engineer.

  Eng Rec (1898) 39:105

Smith, George C. (1838?-95). Railroad engineer,
  Midwest and Uruguay.

  Eng Rec (1894-5) 31:237
  Nat Car (1895) 26:42
  RR Gaz (1895) 27:143

Smith, Gustavus Woodson (1822-96). Civil and military
  engineer; professor at West Point.

  Eng & Min J (1896) 61:619
  Eng Rec (1896) 34:81

  Appleton's Cyc 5:566
  DAB 17:272-3

Smith, Hamilton (1840-1900). Mining and hydraulic
    engineer and consultant.

   Am Manuf (1900) 67:32
   Eng & Min J (1900) 70:16,34*
   Eng Rec (1900) 42:18
   Iron Age (7/12/1900) 66:22

   DAB 17:273-4

Smith, Harrison J. (1854-96). Machinist and
    electrical expert.

   Elec Eng (1896) 21:678*
   Eng Rec (1896) 34:41

Smith, Henry B. (1817?-1900). Manufacturer of steam
    heating apparatus.

   Iron Age (7/12/1900) 66:22

Smith, Horace (1808-93). Co-inventor and manufacturer,
    Winchester rifle and Smith and Wesson revolver.

   Am Eng & RR J (1893) 67:101

   DAB 17:282-3

Smith, Isaac Williams (1826-97). Civil engineer,
    railroads and waterworks.

   ASCE Proc (1897) 23:422-7
   ASCE Trans (1897) 38:456-61

Smith, Israel (? -1879). Chief engineer, New Jersey
    Railroad.

   Eng News (1879) 6:50

Smith, J. Condit (1840-83). Civil engineer.

   Eng News (1883) 10:560
   RR Gaz (1883) 15:761

Smith, James F. (1814?-98). Civil engineer.

   RR Gaz (1898) 30:87

Smith, James Foster (1813-98). Civil engineer, early
    railroads and canals in Pa.

   AISA Bull (1898) 32:28
   Eng & Min J (1898) 65:168
   Eng Rec (1897-8) 37:205
   Frank Inst J (1898) 145:468-71*
   Iron Age (2/3/1898) 61:26

Smith, James G. (1836-1900). Inventor, known for
duplex system of telegraphy.

Elec World (1900) 35:456
Sci Am (1900) 82:179

Smith, James N. (1827?-88). Military engineer and
contractor, railroads.

Eng Rec (1888) 18:130
RR Gaz (1880) 20:512

Smith, John B. (1815-95). Mechanical draftsman;
president, Erie & Wyoming Valley Railroad.

RR Gaz (1895) 27:60

Smith, John Rollins (1839-93). Civil engineer and
proprietor of Springfield Iron Works, specializing
in architectural iron and bridge work.

Iron Age (1893) 52:158

Smith, Lucius A. (1831-94). Marine engineer;
supervised construction of Monitor and other
iron vessels; superintendent of Continental
Iron Works.

ASCE Proc (1894) 20:48-9
Eng (1894) 27:17
Eng Rec (1893-4) 29:102

Smith, Ralphael Palmer (? -1894). Civil engineer,
mining in Calif., hydraulic works in Santo
Domingo.

Eng & Min J (1894) 57:180
Eng Rec (1893-4) 29:200

Smith, Samuel (1815?-88). Manufacturer of locomotives;
boiler maker, Paterson, N.J.

Eng & Min J (1888) 45:238
Iron Age (1888) 41:576

Smith, W. Harrold (? -1885). Mechanical engineer.

Ind World (11/26/1885) 25:6

Smith, Walter W. (1850-96). Mechanical engineer;
designer and builder of hydraulic machinery.

ASME Trans (1897) 18:1091

Smith, Warren Collier (1866-95). Railway engineer.

Assoc Eng Soc J (1895) 15:24-5

Smith, William (1827-92). Locomotive engineer and
master mechanic.

Am Eng & RR J (1892) 66:146
Nat Car (1892) 23:31
RR Gaz (1892) 24:69

Smith, William H. (? -1897). Civil engineer and
surveyor, canals and railroads.

RR Gaz (1897) 29:121

Smith, William Henry (? -1897). Railroad engineer.

Eng Rec (1896-7) 35:223

Smith, William S. (? -1897). Naval engineer.

Am Mach (1897) 20:141

Smyser, Henry E. (? -1899). Mechanical engineer;
inventor of coffee packing and sugar packing
machinery.

Iron Age (3/16/1899) 63:19

Smythe, Andrew E. (1847-97). Gas engineer.

Iron Age (7/15/1897) 60:18

Smythe, Samuel R. (1860-99). Metallurgical engineer;
builder of large open-hearth steel works and
gas-producing plant.

Am Manuf (1899) 65:387
Eng & Min J (1899) 68:585
Iron Age (11/2/1899) 64:20

Snow, Chauncey H. (1833?-93). Member, Corps of
Engineers; worked on Hoosac Tunnel.

Eng & Min J (1893) 55:492
Eng Rec (1892-3) 27:506

Snow, William B. (1821-98). Master mechanic,
Illinois Central Railroad.

Am Eng & RR J (1898) 72:418
Am Mach (1898) 21:829
RR Gaz (1898) 30:785

Snyder, George W. (1805-86). Machinist and manufacturer of machinery for coal mining.

AISA Bull (1886) 20:291
RR Gaz (1886) 18:695

Snyder, John H. (? -1892). Inventor of railroad devices; superintendent of Albany Iron Works.

AISA Bull (1892) 26:201
Eng & Min J (1892) 54:36
Eng Rec (1892) 26:70
Iron Age (1892) 50:19

Spatswood, George (1846-96). Mining engineer.

Eng & Min J (1896) 62:491

Spaulding, Ira (1818-75). Engineer, railroads.

ASCE Proc (1873-5) 1:338-40
RR Gaz (1875) 7:419

Speakman, Thomas Say (1817?-97). Inventor of mining equipment.

Eng & Min J (1897) 63:215
Eng Rec (1896-7) 35:267

Specht, George D. (? -1888). Civil engineer.

Am Eng & RR J (1888) 62:572

Speilmann, Arthur (1847-83). Civil engineer, specialist in water supply; professor at N. Y. University.

ASCE Proc (1884) 10:115-7
Eng News (1883) 10:587

Spooner, D. Brainerd (1830-90). Inventor of water meter.

Eng & Min J (1890) 49:645
Iron Age (1890) 45:1002

Sprague, Adelbert L. (1872-95). Surveyor.

Assoc Eng Soc J (1895) 14:11-2

Stace, Arthur J. (1838-90). Civil engineer, author and professor at Notre Dame.

Am Eng & RR J (1890) 64:522
Eng Rec (1890) 22:258
Iron Age (1890) 46:534

Stager, Anson (1825-85). Pioneer builder and operator of telegraphs; originated military telegraph cipher system; president of Western Electric.

Elec Eng (1885) 4:157-8
Ind World (4/2/1884) 24:7
Sanit Eng (1884-5) 11:378

DAB 17:492-3
NYT (3/23/1885) 1:2

Stahlberg, Albert Jacob (1846-87). Civil engineer, waterworks, railways and city improvements.

ASCE Proc (1896) 22:618
ASCE Trans (1896) 36:597

Standish, David B. (1817?-80). Locomotive engineer.

RR Gaz (1880) 12:133

Stanton, Isaac (1813?-1900). Builder of steam engines and machinery.

Am Mach (1900) 23:1222
Iron Age (12/20/1900) 66:32

Stanwood, James Hugh (1860-96). Civil engineer; instructor at MIT.

ASCE Proc (1896) 22:155
ASCE Trans (1896) 36:590
Assoc Eng Soc J (1896) 17:23-4
Eng & Min J (1896) 61:523
Eng Rec (1896) 33:453
RR Gaz (1896) 28:384

Stark, George (1822?-92). Civil engineer, railroads.

Am Eng & RR J (1892) 66:240
Eng Rec (1891-2) 25:326

Starling, William (? -1900). Civil engineer.

Eng Rec (1900) 42:579

Stearns, Joseph Barker (1831-95). Inventor of a duplex system of telegraphy.

Elec Eng (1895) 20:37*

Stearns, William B. (1826?-83). Civil engineer.

RR Gaz (1883) 15:594

Stearns, William Ellison (1857-98). Mechanical engineer.

ASME Trans (1898) 19:977

Stearns, William H. (1822?-99). Master mechanic, railroads in Connecticut.

RR Gaz (1899) 31:530

Stedman, A. W. (1844?-98). Chief civil engineer, Lehigh Valley Railroad.

Am Eng & RR J (1898) 72:384
Eng Rec (1898) 38:421
RR Gaz (1898) 30:749

Steele, J. Dutton (1810-86). Civil engineer, canals and railroads; made improvements in bridges.

Eng & Min J (1896) 62:559-60
RR Gaz (1886) 18:432

Steinbecher, Charles (? -1886). Marine engineer.

Eng (1886) 11:98

Steinmetz, William (? -1898). Civil engineer and architect.

Eng Rec (1897-8) 37:513

Stephenson, John (1809-93). Inventor and builder of first streetcar.

Am Eng & RR J (1893) 67:453
Eng & Min J (1893) 56:144
Eng Rec (1893) 28:150
Nat Car (1893) 24:148
RR Gaz (1893) 25:596
Sci Am (1893) 69:98

DAB 17:583-4

Stetefeldt, Carl August (1838-96). Inventor and metallurgist; developed furnace for treating ores containing gold and silver.

AIME Trans (1896) 26:537-44
Eng & Min J (1896) 61:300*

Appleton's Cyc 5:668
DAB 17:595

Stevens, Alonzo J. (1828-1900). Manufacturer of mill gearing and shafting, iron and brass castings and woodworking machinery.

Iron Age (9/6/1900) 66:21

Stevens, Andrew J. (1833-88). Master mechanic and locomotive engineer.

Am Eng & RR J (1888) 62:190
Am Mach (3/10/1888) 11:8
Nat Car (1888) 19:41
RR Gaz (1888) 20:145

Stevens, Edwin Augustus (1795-1868). Inventor of improvements in marine and other machinery; improved methods of ironplating.

Sci Am (1868) 16:136

DAB 17:608-9
NYT (8/11/1868) 8:2

Stevens, Everett A. (1843?-95). Locomotive engineer and railroad commissioner.

RR Gaz (1895) 27:488

Stevens, George W. (? -1897). Civil engineer and architect.

Eng Rec (1897) 36:399

Stevens, John (1749-1838). Engineer and inventor; pioneer in the field of mechanical transportation; developed screw propeller, steam engine, boilers.

Pop Sci (1877-8) 12:264
Sci Am (1873) 28:274

Appleton's Cyc 5:673-4*
DAB 17:614-6

Stevens, John G. (? -1885). Consulting engineer, mining and canal works.

RR Gaz (1886) 18:29

Stevenson, William (? -1894). Civil engineer, railroads.

Eng Rec (1893-4) 29:248
RR Gaz (1894) 26:202

Stewart, Charles B. (? -1881). Civil engineer.

Sci Am (1881) 44:49

Stewart, Neil (? -1891). Civil engineer.

Eng Rec (1890-1) 23:289

Stillman, Thomas B. (1806-66). Mechanical engineer, steamships.

Sci Am (1866) 14:35

Appleton's Cyc 5:690

Stiness, Samuel G. (? -1894). Gas engineer.

Eng Rec (1894) 30:388

Stockwell, Emory (? -1896). Inventor of improvements in locks for safes.

Am Mach (1896) 19:211

Stoelting, Herman (? -1875). Metallurgist.

Eng & Min J (1875) 19:348

Stone, Andros B. (1826?-96). Iron manufacturer and bridge builder.

AISA Bull (1896) 30:285
Eng & Min J (1896) 62:587
Iron Age (1896) 58:1265
RR Gaz (1896) 28:906

Stone, Charles Pomeroy (1824-87). Military engineer; constructing engineer for foundations of Statue of Liberty.

Am Eng & RR J (1887) 61:53
Eng & Min J (1887) 43:81
Sanit Eng (1886-7) 15:213

DAB 18:72

Stone, G. H. (1819-79). Civil engineer.

Eng News (1879) 6:385

Stone, Lemuel M. E. (1820?-95). Commissioner of dams and reservoirs in Rhode Island; railroad builder.

Eng Rec (1894-5) 31:291

Stone, Waterman (1847-96). Civil and mechanical engineer; worked on rapid transit.

ASCE Proc (1899) 25:327
ASCE Trans (1899) 41:649
Eng Rec (1895-6) 33:309
RR Gaz (1896) 28:241

Strange, John S. (? -1894). Civil engineer and surveyor, New Jersey.

Eng Rec (1894) 30:336

Stratton, Alexander (1872-99). Electrical engineer.

AIEE Trans (1899) 16:696

Stratton, Franklin A. (1829-79). Civil engineer; worked on several railroads and at League Island Navy Yard.

ASCE Proc (1879) 5:93-6
Eng News (1879) 6:233

Street, C. B. (1838?-90). Master mechanic, railroads.

RR Gaz (1890) 22:847

Striedinger, Julius Hermann (1841-94). Civil engineer, mining and hydraulic works; worked primarily in Latin America.

ASCE Proc (1894) 20:198-200
Eng & Min J (1894) 58:395
Eng Rec (1894) 30:354
RR Gaz (1894) 26:746

Strong, W. W. (? -1894). Civil engineer.

Eng Rec (1894) 30:166

Stroud, Reuben W. (1841?-75). Civil engineer, railroads and canals in New York State.

RR Gaz (1875) 7:509

NYT (12/3/1875) 4:7

Stuart, Charles Beebe (1814-81). Railroad engineer in N.Y. State; later involved in dock construction and naval architecture.

RR Gaz (1881) 13:7

Appleton's Cyc 5:728
DAB 18:163

Sturtevant, Benjamin Franklin (1833-90). Inventor and manufacturer of wood-handling, shoe-pegging and other machines.

Eng & Min J (1890) 49:453
Eng Rec (1889-90) 21:322
Iron Age (1890) 45:688

DAB 18:184-5

Stutz, Sebastian (1833?-1900). Mining and mechanical engineer; inventor of mining appliances.

Am Manuf (1900) 67:51
Iron Age (7/19/1900) 66:16

Stwolinski, Ferdinand de (1839-92). Mining engineer.

Eng & Min J (1891) 52:501*; (1892) 53:432

Sublett, David L. (1837?-96). Civil engineer and geologist.

Eng & Min J (1896) 61:307

Sullivan, John (? -1889). Engineer and contractor, railroads and mines.

Eng Rec (1888-9) 19:102

Summers, Charles H. (? -1898). Chief electrician, Western Union.

Elec World (1898) 32:518
RR Gaz (1898) 30:819

Swan, Charles Herbert (1842-99). Civil engineer, sewer and water works.

Assoc Eng Soc J (1899) 23:6-8*

Sweet, Arthur R. (? -1898). Civil engineer; town engineer at Pawtucket and Woonsocket, R.I.

Eng Rec (1898) 38:465

Sweet, Homer B. (1826?-93). Civil engineer; surveyor in Adirondacks.

Eng & Min J (1893) 56:526

Sweet, Sylvanus Howe (1830?-99). Civil engineer, state canals in N.Y.

Eng Rec (1899) 40:611
RR Gaz (1890) 31:318

Swett, James H. (1814?-80). Inventor of machine for
producing railroad spikes; manufacturer of spikes.

Am Manuf (4/16/1880) 26:10

Swift, McRee (1819-96). Civil engineer; manufacturer
of lead pipe in N.Y.; water commissioner, New
Brunswick.

ASCE Trans (1896) 36:616-7
Eng Rec (1895-6) 33:327

Swift, William Henry (1800-79). Civil and topographical
engineer, railroads; soldier.

RR Gaz (1879) 11:216
Sci Am (1879) 40:354

DAB 18:249-50

Swoyer, John Henry (1872-99). Mining engineer.

AIME Trans (1900) 30:xxxviii

Sykes, Lorenzo A. (1806-78). Civil engineer, canals,
railroads and public works.

Eng News (1878) 5:406
RR Gaz (1878) 10:604

Symington, William Newton (1841-99). Mining engineer
and metallurgist.

AIME Trans (1900) 30:xxxviii

Symons, William R. (1824?-88). Mining engineer.

Eng & Min J (1888) 45:112

Taber, Henry S. (? -1894). Military engineer;
responsible for river and harbor improvements in
Middle West.

Eng Rec (1893-4) 29:328

Taintor, William Noyes (1870-98). Civil engineer,
hydraulic and sanitary works.

ASCE Trans (1898) 39:704

Talcott, Andrew (1797-1883). Military engineer, several fortifications; civil engineer, railroads.

Eng News (1883) 10:200
RR Gaz (1883) 15:271

DAB 18:281

Talcott, Cook (1826-93). Civil engineer, canals and tunnels; chief engineer, St. Louis.

ASCE Proc (1893) 19:99-101
Eng Rec (1892-3) 27:151

Talcott, Edward Benton (1812-86). Civil engineer, canal, railroad and harbor works.

Assoc Eng Soc J (1885-6) 5:364-7
RR Gaz (1886) 18:136,152

Talcott, George Russell (? -1899). Civil and railway engineer.

Eng Rec (1898-9) 39:291

Talcott, William H. (1809-68). Civil engineer, canals and hydraulic works.

ASCE Proc (1893) 19:97-9
Van Nostrand's (1869) 1:31

Tasker, Charles A. (? -1879). Civil engineer, railroads.

ASCE Proc (1881) 7:91

Tasker, Stephen P. M. (1835?-1900). Inventor, mechanical engineer and manufacturer.

Am Mach (1900) 23:305
Iron Age (3/22/1900) 65:20

Taylor, Archibald R. (? -1891). Civil engineer.

Am Eng & RR J (1891) 65:524

Taylor, Charles M. (1817?-93). Civil engineer, canals and mines in Ohio.

Eng & Min J (1893) 56:454

Taylor, J. Archie (1857-92). Manufacturer of marine and general machinery.

ASME Trans (1892) 13:684

216

Taylor, William B. (? -1895). Civil engineer and surveyor; New York State engineer.

Eng Rec (1894-5) 31:183

Temple, John Frederic (1865-95). Civil engineer, railroads; bridge draftsman.

ASCE Proc (1895) 21:183

Thatcher, Horace C. (? -1898). Submarine engineer and contractor.

Eng Rec (1897-8) 37:271

Thaxter, George E. (? -1890). Inventor of electric door lock.

Elec World (1890) 15:339

Thayer, Sylvanus (1785-1872). Military engineer; designer and builder of fortifications.

RR Gaz (1872) 4:413

DAB 18:410-1

Thayer, William W. (1802-93). Civil engineer, railroads; inspector of mines and bridges; responsible for tunnels and parks in Philadelphia.

ASCE Proc (1893) 19:94-7
Eng Rec (1892-3) 27:311

Thayer, Winthrop (1863?-1900). Electrical and steam engineer.

Am Mach (1900) 23:728
Elec World (1900) 36:152

Theall, Horace (1821-96). Manufacturer of marine engines and boilers.

Iron Age (1896) 58:413

Thomas, Arthur T. (? -1900). City engineer of Superior, Wisconsin.

Eng Rec (1900) 42:453

Thomas, David (1794-1882).   Pioneer iron manufacturer;
    developed process of smelting iron with anthracite
    coal.

    AISA Bull (1882) 16:473
    Am Manuf (6/30/1882) 30:1; (10/6/1882) 31:11
    Eng & Min J (1882) 33:325
    Iron Age (11/9/1876) 18:3; (6/22/1882) 29:15-6*
    RR Gaz (1882) 14:332
    Sci Am (1882) 47:3

    Appleton's Cyc 6:78
    DAB 18:427-8

Thomas, John (1829-97).   Iron manufacturer, eastern Pa.

    Am Manuf (1897) 60:444
    ASME Trans (1898) 19:965
    Eng & Min J (1897) 63:310
    Eng Rec (1896-7) 35:355
    Iron Age (3/25/1897) 59:18

Thomas, Joseph Russell (1820-96).   Gas engineer and
    technical editor.

    ASCE Proc (1897) 23:327-8
    ASCE Trans (1897) 37:566-8
    Eng Rec (1896-7) 35:3

Thomas, Miner T. (? -1897).   Civil engineer, railroads.

    Eng Rec (1897) 36:421

Thomas, William Knapp (? -1897).   Mechanical engineer,
    assisted in construction of Monitor.

    Eng Rec (1896-7) 35:377
    Iron Age (4/1/1897) 59:15

Thomes, John E. (? -1893).   Civil engineer and
    contractor.

    Eng Rec (1892-3) 27:311

Thompson, George L. (? -1895).   Civil engineer.

    Eng Rec (1395-6) 33:93

Thompson, Isaac F. (1837?-98).   Mining engineer and
    inventor, Pacific coast; developed steam cut-off
    valve engine.

    Eng & Min J (1898) 65:78
    Eng Rec (1897-8) 37:139

Thompson, John Chambers (1844-80). Civil engineer, waterworks and railroads.

ASCE Proc (1896) 22:695
ASCE Trans (1896) 36:584

Thompson, John Edgar (1808-74). Civil engineer, railroads; president of Pennsylvania Railroad.

Iron Age (6/4/1874) 13:5,14-5

NYT (5/29/1874) 4:6

Thompson, John J. (1816?-75). Manufacturer of machinery.

AISA Bull (1875) 9:164

Thompson, John Polk (1838?-99). Inventor of cotton-mill machinery.

Am Mach (1899) 22:923

Thompson, William T. (1857-99). Mining engineer, western North America.

AIME Trans (1900) 30:xxxviii

Thomson, Frank (1841-99). Mechanical engineer, railroads; made improvements in locomotives, track maintenance, signals; president, Pennsylvania Railroad.

Am Eng & RR J (1899) 73:232
AISA Bull (1897) 31:41; (1899) 33:106
Am Mach (1899) 22:545
Eng & Min J (1899) 22:545
Eng Rec (1899) 40:25
Iron Age (6/8/1899) 63:18
RR Gaz (1899) 31:399*
Sci Am (1899) 80:389

DAB 18:483-4

Thurston, Robert Lawton (1800-74). Manufacturer and builder of steam engines.

Eng (1890) 19:31*
Iron Age (1/29/1874) 13:16
Sci Am (1874) 30:145

DAB 18:520-1

Tidd, Marshall Martain (1827-95). Civil and
hydraulic engineer.

ASCE Proc (1897) 23:171-3
ASCE Trans (1897) 37:568
Assoc Eng Soc J (1897) 18:66-8
Eng & Min J (1895) 60:179
Eng Rec (1895) 32:219,237

Tilghman, Richard Albert (1824-99). Developer of
sand-blasting and sand-blasting machinery.

Am Mach (1899) 22:268
Eng & Min J (1899) 67:416
Iron Age (1/30/1899) 63:18

DAB 18:544-5

Tinsley, A. J. (? -1895). Naval engineer; space
manager of telegraph office.

Elec Eng (1895) 20:556

Tobey, William Boardman (1869?-96). Electrical
engineer.

Elec Eng (1896) 21:373
Elec World (1896) 27:446

Torrence, Joseph Thatcher (1843-96). Iron and
steel manufacturer; builder of railroads,
Chicago.

Am Eng & RR J (1896) 70:335
Eng & Min J (1896) 62:443
Eng Rec (1896) 34:421
Iron Age (1896) 58:871

DAB 18:594

Totten, George Muirson (1809-84). Chief engineer in
construction of Panama Railroad.

Eng News (1884) 12:259
Iron Age (5/22/1884) 33:17
RR Gaz (1884) 16:401

Appleton's Cyc 6:140-1
DAB 18:598
NYT (5/20/1884) 5:2

Totten, Joseph Gilbert (1788-1864). Military engineer;
developed system of lighting by Fresnel lenses.

ASCE Proc (1896) 22:681
ASCE Trans (1896) 36:525-7

Appleton's Cyc 6:141*
DAB 18:598-9

Tower, George B. N. (? -1889). Civil engineer;
chief engineer, U. S. Navy.

Eng & Min J (1889) 48:297

Tower, Zealous B. (1819-1900). Military engineer,
river, harbor and fortification work.

Eng Rec (1900) 41:281

Appleton's Cyc 6:146

Towle, Hamilton E. (? -1881). Mechanical engineer.

Sci Am (1881) 45:272

Towne, Lewis W. (? -1892). Locomotive engineer,
and superintendent of railroads.

Am Eng & RR J (1892) 66:337
Nat Car (1892) 23:94

Tracy, Edward H. (1817-75). Chief engineer, Croton
Aqueduct; canal, railroad and dock work.

ASCE Proc (1873-5) 1:337-8
RR Gaz (1875) 7:368
Sci Am (1875) 33:180

Trafton, Gilman (? -1887). Bridge engineer; helped
form Louisville Bridge and Iron Company.

Am Eng & RR J (1887) 61:298
ASCE Proc (1889) 15:42-3

Trautwine, John Cresson (1810-83). Civil engineer and
railroad builder; chief engineer, Panama Railroad;
writer of manuals.

Cassier's (1893-4) 5:233-6*
Eng News (1883) 10:450
Iron Age (9/27/1883) 32:15
RR Gaz (1883) 15:624,626,647*
Sci Am (1883) 49:210

Appleton's Cyc 6:154-5
DAB 18:628-9
NYT (9/16/1883) 2:6

Treadwell, Daniel (1791-1872). Inventor of screw and nail making machinery, printing press and other machines.

Sci Am (1872) 26:182

Appleton's Cyc 6:155
DAB 18:631-3
NYT (3/2/1872) 1:6

Trik, Carl (? -1898). Bridge builder; superintendent of bridges, Philadelphia.

Eng Rec (1898) 38:135

Tripp, Seth D. (? -1898). Inventor of improvements in shoe manufacturing machinery.

Am Mach (1898) 21:38

Trowbridge, William Petit (1828-92). Professor of engineering, Columbia School of Mines; important in development of engineering education.

Am Eng & RR J (1892) 66:432
Am Mach (8/25/1892) 15:8
ASME Trans (1892) 13:684
Eng & Min J (1892) 54:197*
Eng Rec (1892) 26:180
RR Gaz (1892) 24:624

DAB 18:656-7

Truesdell, Charles (1833-94). Civil engineer, water works in Syracuse.

ASCE Proc (1896) 22:693-4
ASCE Trans (1896) 36:585-6
Eng Rec (1893-4) 29:345

Tucker, James (? -1898). Locomotive engineer.

RR Gaz (1898) 30:366

Tufts, Otis (? -1869). Inventor; improver of elevator, steam engine; builder of iron steamer.

Sci Am (1869) 21:345

Turreff, W. F. (1834-92). Superintendent of motive power; railroads; master mechanic and car builder.

Am Eng & RR J (1892) 66:96
Nat Car (1892) 23:31

Turtle, Thomas (? -1894). Military engineer.

   Eng & Min J (1894) 58:276
   Eng Rec (1894) 30:266
   RR Gaz (1894) 26:656

Tweddell, Ralph Hart (1842-95). Developer of hydraulic tools.

   ASME Trans (1895) 16:1198

Tweedale, William (1823-1900). Civil engineer, bridges and buildings.

   Eng Rec (1900) 42:453

   Appleton's Cyc 6:190

Tynan, J. W. (1837-92). Naval engineer.

   ASME Trans (1892) 13:680-1

Tyson, Henry (1820-77). Engineer, railroads; engine builder.

   ASCE Proc (1878) 4:110-2
   RR Gaz (1877) 9:407

Uhlinger, W. P. (1823-98). Manufacturer of Jacquard machine and ribbon loom.

   Am Mach (1898) 21:395-6
   Iron Age (5/12/1898) 61:28

Underhill, Arthur B. (1832-96). Superintendent of motive power, railroads.

   Am Eng & RR J (1896) 70:147

Vaillant, George Amedie (? -1896). Civil engineer.

   Eng Rec (1896) 34:197

Van Auden, William (1815?-92). Inventor of labor saving devices.

   Eng & Min J (1892) 53:574
   Iron Age (1892) 49:1037

Van Brocklin, Martin (? -1896). Civil engineer,
railroads in New York, Peru, Australia, and
Nebraska.

Eng Rec (1895-6) 33:255
Iron Age (1896) 57:707
RR Gaz (1896) 28:187

Van Buren, Daniel T. (1825?-90). Military engineer
and surveyor.

Eng & Min J (1890) 50:80
Eng Rec (1890) 22:98

Van Depoele, Charles J. (1846-92). Electrical
engineer, pioneer in electric light and traction;
developer of electric street railways.

Am Eng & RR J (1892) 66:192
AIEE Trans (1892) 9:175-8
Elec Eng (1890) 9:495-6*; (1892) 13:317
Elec World (1892) 19:210*
Eng Rec (1891-2) 25:274
Iron Age (1892) 49:564
RR Gaz (1892) 24:235
Sci Am (1892) 66:217*

DAB 19:168-9

Vander Weyde, P. H. (? -1895). Inventor of electrical
devices and editor of Manufacturer and Builder.

Elec Eng (1895) 19:296
Eng & Min J (1895) 59:275
Sci Am (1895) 72:197

Van Liew, Evander S. (? -1897). Civil engineer.

Eng Rec (1897) 36:465

Van Nortwick, John (? -1890). Engineer, railroads.

Am Eng & RR J (1890) 64:281

Van Nostrand, David (1811-86). Publisher of scientific
and engineering magazines and texts.

Eng & Min J (1886) 41:445
Sanit Eng (1886) 14:63
Van Nostrand's (1886) 35:441-5

Appleton's Cyc 6:249-50
DAB 19:203-4

Van Rensselaer, Bernard (? -1879). State engineer and
surveyor.

Eng News (1879) 6:210

Van Rensselaer, Jeremiah (1812-74). Railroad
engineer.

RR Gaz (1874) 6:273

Vanderbilt, Cornelius H. (1794-1877). Steamboat and
railroad magnate.

Am Eng & RR J (1899) 73:314
Iron Age (1/11/1877) 19:5
Sci Am (1877) 36:37

Appleton's Cyc 6:240-1
DAB 19:169-73
NYT (1/5/1877) 1:5

Varney, Enoch (1812?-80). Master railroad car builder.

Nat Car (1881) 12:178

Vaughan, Frederic W. (1844-87). Civil engineer,
bridges and railroads; president, Louisville Bridge
and Iron Company.

Am Eng & RR J (1887) 61:495
ASCE Proc (1889) 15:44-6
RR Gaz (1887) 19:724
Sanit Eng (1887) 16:588

Veazie, Joseph (? -1897). Civil engineer, Mass.

Eng Rec (1896-7) 35:333

Veeder, John Irwin (1858-92). Mechanical engineer.

ASME Trans (1894) 15:1196-8

Venable, W. W. (? -1888). Mechanical engineer,
New Hampshire.

Power (11/1888) 8:11

Verbryck, Benjamin (? -1891). Master car builder.

Nat Car (1891) 22:109*

Vernon, George (1807?-92). Locomotive engineer.

RR Gaz (1892) 24:676

Vickers, Albert (1868-98). Electrical engineer,
contributor to electrical journals.

Elec World (1898) 32:127

Viebahn, Gustav O. (? -1897). City engineer, Watertown.

Eng Rec (1896-7) 35:223

Vinton, Francis Laurens (1835-79). Military and
mining engineer; one of organizers of Columbia
School of Mines.

ASCE Proc (1882) 8:91
Eng & Min J (1879) 28:257-8
Eng News (1879) 6:326

DAB 19:282-3
NYT (10/7/1879) 2:4

Vliet, William (? -1893). Civil engineer; built
first bridge across the Mississippi at Minneapolis.

Eng Rec (1892-3) 27:333

Von der Bosch, William (? -1893). Engineer, employed
in construction of East River Bridge.

Eng Rec (1893) 28:246

Voorhees, Philip R. (1835-95). Mechanical engineer,
ships.

ASME Trans (1896) 17:741

Vose, Richard (1830-93). Manufacturer of and inventor
of improvements in car springs.

Am Eng & RR J (1893) 67:201
Eng & Min J (1893) 55:204
RR Gaz (1893) 25:178

Wade, Jephtha Homer (1811-90). Telegrapher; one of
the founders of the American commercial telegraph
system; a president of Western Union.

Elec Eng (1890) 10:171
Elec World (1890) 16:107

DAB 19:306-7

Wadsworth, Alexander (? -1898). Civil engineer,
waterworks.

Eng Rec (1897-8) 37:271

Wagner, John R. (1860?-99). Mechanical engineer.

AIME Trans (1900) 30:xxxix
ASME Trans (1899) 20:1007

Wait, George W. (? -1894). Marine engineer.

Eng Rec (1894) 30:182

Waite, Christopher Champlin (1844?-96). Engineer; president of Columbus, Hocking Valley & Toledo Railroad.

Eng Rec (1895-6) 33:219
RR Gaz (1896) 28:152

Walbridge, A. C. (? -1892). Engineer and builder, Coney Island & Brooklyn Railroad.

Eng Rec (1892) 26:194

Walcott, Benjamin S. (1786-1862). Manufacturer of machinery for handling cotton.

Sci Am (1862) 6:74

Walker, George (? -1892). Civil engineer and surveyor.

Eng & Min J (1892) 53:113

Walker, William Weightman (1865-91). Mining engineer.

Eng & Min J (1891) 51:612

Walker, William Williams (1834-93). Civil engineer, railroads in Midwest; later involved with newspapers.

ASCE Proc (1893) 19:174-5

Walker, Zephon F. (? -1897). Civil engineer and surveyor.

Eng Rec (1897) 36:465

Wall, Edward Barry (1856-94). Mechanical engineer, railroads.

Am Eng & RR J (1894) 68:231-4*
ASME Trans (1894) 15:1190-1
Eng Rec (1893-4) 29:296
Nat Car (1894) 25:76
RR Gaz (1894) 26:256

Wallace, Edward R. (? -1893). Civil engineer.

Eng Rec (1893) 28:326

Wallace, William (1805-87). Civil engineer, railroads in New York.

Am Eng & RR J (1887) 61:348
RR Gaz (1887) 19:426
Sanit Eng (1887) 16:100

227

Waller, William Gustavus (1813-91). Civil engineer
and surveyor, railroads.

Am Eng & RR J (1891) 65:381
Eng Rec (1891) 24:52

Walling, Henry Francis (1825-89). Surveyor; professor
of topographical and civil engineering, Lafayette
College; mapmaker for U.S.G.S.

ASCE Proc (1889) 15:140-1
Assoc Eng Soc J (1889) 8:28-30

Walton, Louis Roberts (1842-85). Civil engineer,
harbors and mines.

ASCE Trans (1896) 36:586-7

Walworth, James J. (? -1896). Developer of steam and
hot water apparatus for heating and ventilating.

Eng Rec (1895-6) 33:380

Wanich, Alexander (1825?-96). Mechanical engineer.

Iron Age (1896) 58:587

Ward, Peter (? -1891). Railroad engineer and
contractor.

Am Eng & RR J (1891) 65:283
Eng Rec (1890-1) 23:388
RR Gaz (1891) 23:343

Ward, William E. (1821-1900). Inventor of machinery
for producing hardware; manufacturer of bolts,
nuts and wood screws.

Am Mach (1900) 23:236
ASME Trans (1900) 21:1164-5
Eng Rec (1900) 41:233
Iron Age (3/8/1900) 65:19

Wardlow, James Robert (1856-91). Civil and mining
engineer.

ASCE Proc (1892) 18:96-7

Waring, George Edwin (1833-98). Sanitary engineer;
worked on alleviation of yellow fever conditions.

AISA Bull (1898) 32:172
Eng News (1898) 40:284*,296
Eng Rec (1898) 38:501
Iron Age (11/3/1898) 61:18-9
RR Gaz (1898) 30:803
Sci Am (1898) 79:291

Appleton's Cyc 6:358-9
DAB 19:456-7
NYT (10/30/1898) 1:7,2:1

Warner, James Cartwright (1830-95). Manufacturer of
pig iron; developer of iron industry in Tenn. and Ga.

AISA Bull (1883) 17:19; (1895) 29:171
Iron Age (1895) 56:286

DAB 19:465-6

Washburn, Ichabod (1798-1868). Manufacturer of machinery
for carding and spinning wool; manufacturer of wire.

Ind World (7/26/1888) 31:6-7
Sci Am (1869) 20:43
Van Nostrand's (1869) 1:102

DAB 19:501-2

Washburn, Thurlow (1869?-99). Mining engineer and
assayer.

Eng & Min J (1899) 68:524

Waterman, H. E. (? -1898). Member, Army Corps of
Engineers; worked on river and harbor improvements.

Eng Rec (1898) 38:465

Waters, Thomas J. (? -1898). Mining engineer,
American West and Far East.

AIME Trans (1899) 29:xxxvi-vii
Eng & Min J (1898) 65:228
Eng Rec (1897-8) 37:249

Watson, Frederick M. (1865-1900). Mining engineer,
South Africa and Latin America.

Eng & Min J (1900) 69:236,326
Iron Age (2/22/1900) 65:21

Watson, Peter H. (1817?-85). Civil engineer.

RR Gaz (1885) 17:492

Watson, Thomas H. (1836-96). Machinist and
manufacturer of hydraulic machinery.

Am Mach (1896) 19:442
Eng & Min J (1897) 64:162
Iron Age (1896) 57:875

Weaver, Charles B. (? -1895). Engineer, waterworks.

Eng Rec (1895-6) 33:21

Weaver, Norman Rutherford (1869-97). Steam and
electrical engineer.

ASME Trans (1898) 19:973-4

Webb, George (1828-83). Civil engineer, railroads
and steel industry.

AISA Bull (1883) 17:289
Eng & Min J (1883) 36:248
RR Gaz (1883) 15:679

Webb, William H. (1816-99). Shipbuilder.

Am Mach (1899) 22:1074

Appleton's Cyc 6:404
DAB 19:578-9

Webster, Daniel Atley (? -1881). Civil engineer;
inventor of planing machines, street-sweeping
machine, and other devices.

Ind World (3/10/1881) 16:18
Sci Am (1881) 44:216

Webster, John Fraser (1848-93). Designer of machines.

Am Mach (2/2/1892) 16:8
ASME Trans (1893) 14:1443

Webster, John H. (1850-95). Designer of machinery;
operator of sugar refinery, Boston.

ASME Trans (1895) 16:1196
Assoc Eng Soc J (1895) 15:20
Eng & Min J (1895) 59:347

Wederkinch, Carl O. (? -1881). Civil engineer,
railroads.

RR Gaz (1881) 13:390

Wedge, Francis (1825-93). Machinist and manufacturer
of machinery.

Iron Age (1893) 51:792

Weekes, John (1797?-1891). Railroad engineer, New
  York Central.

  RR Gaz (1891) 23:293

Weeks, Joseph Dame (1840-96). Mining engineer; a
  president of the AIME; editor of American
  Manufacturer.

  AIME Trans (1897) 27:231-8
  AISA Bull (1897) 31:5,9
  Am Manuf (1897) 60:7-8*,91
  Eng & Min J (1897) 63:47
  Eng News (1896) 35:34-5
  Ind World (12/3/1896) 48:5
  Iron Age (1896) 58:1318
  RR Gaz (1897) 29:17

  DAB 19:602-3

Weimer, P. L. (1830-91). Draftsman and inventor,
  blast furnace machinery.

  Am Manuf (1891) 49:511

Weimer, Thomas (? -1887). Engineer, canals.

  Eng & Min J (1887) 44:100

Weir, Fred C. (1832-99). Mechanic, inventor and
  railroad engineer; manufacturer of frogs, Cincinnati.

  Am Eng & RR J (1899) 73:137
  Am Mach (1899) 22:311
  Eng & Min J (1899) 31:178
  Iron Age (3/16/1899) 63:19

Weir, Silas E. (? -1898). Railroad engineer,
  Kentucky and Pa.; city surveyor in New Brunswick.

  Eng Rec (1898) 38:421

Weirman, Thomas T. (? -1887). Civil engineer;
  surveyor of Brooklyn docks.

  RR Gaz (1887) 19:531

Weitzel, Godfrey (1835-84). Military engineer, river
  and harbor improvements in Delaware, ship canals
  in Midwest.

  Iron Age (3/27/1884) 33:17
  DAB 19:616-7

Welch, Ashbel (1809-82). Civil engineer, railroads
and canals; applied telegraphy to railroad
signals; a president of the ASCE.

ASCE Proc (1883) 9:137-44
Eng & Min J (1882) 34:171
Eng News (1882) 9:342
Sanit Eng (1882) 6:343,366
Sci Am (1882) 47:232

DAB 19:618-9

Weld, Frederick F. (1850-90). City engineer,
Waterbury, Conn.

ASCE Proc (1890) 17:203-5
Eng Rec (1890) 22:66

Weld, Henry Thomas (1816-93). Civil engineer;
manufacturer of rails.

AISA Bull (1893) 27:220
Eng & Min J (1893) 56:84

Wellington, Arthur Mellen (1847-95). Civil and
consulting engineer, railroads in Eastern U.S.
and Mexico; author of books on railway locating.

ASCE Proc (1895) 21:199-202
ASME Trans (1895) 16:1197-8
Assoc Eng Soc J (1895) 15:2-3
Elec Eng (1895) 19:500
Eng & Min J (1895) 59:511*
Eng News (1895) 33:337*
Eng Rec (1894-5) 31:453
Iron Age (1895) 55:1084
Nat Car (1895) 26:90
Power (6/1895) 15:15
RR Gaz (1895) 27:337

Appleton's Cyc 6:428
DAB 19:634

Wells, Bard (? -1893). Mining engineer.

Eng & Min J (1893) 55:540

Wendt, Arthur Frederick (1852-93). Mining engineer;
builder of plant in South America.

ASME Trans (1893) 14:1450
Eng & Min J (1893) 56:393*
Eng Rec (1893) 28:310
Iron Age (1893) 52:665
RR Gaz (1893) 25:756

West, John Gartrell (? -1893). Mechanical engineer.

Eng & Min J (1893) 55:444,468
Iron Age (1893) 51:1128

Westerman, James (1819-84). Coal developer, inventor, and operator of iron works.

AISA Bull (1884) 18:196
Iron Age (7/31/1884) 34:17

Weston, James A. (? -1895). Civil engineer, New Hampshire.

Eng Rec (1894-5) 31:417

Weston, William (? -1889). Military engineer.

Eng Rec (1888-9) 19:320

Wetherill, Samuel (1821-90). Manufacturer of metallic zinc; improved zinc ore reduction.

Eng & Min J (1890) 49:737

Appleton's Cyc 6:445
DAB 20:23-4

Wheeler, Charles Yandes (1843?-99). Inventor of armor piercing projectile; president of Firth Sterling Steel Company.

AISA Bull (1899) 33:157
Am Mach (1899) 22:875
Am Manuf (1899) 65:225*
Eng & Min J (1899) 68:314

Wheeler, Nathaniel (1820-93). Inventor of improvements in sewing machines and ventilating systems; manufacturer of buttons, buckles, etc.

Am Mach (1/11/1894) 17:8

Appleton's Cyc 6:454
DAB 20:52

Wheeler, Norman W. (1829-89). Mechanical engineer; inventor of devices for steam pumps.

Am Mach (10/17/1889) 12:8
Eng (1889) 18:91
Eng & Min J (1889) 48:321
Eng Rec (1889) 20:268

Wheeler, Orlando Belina (1835-96). Civil engineer.

ASCE Trans (1896) 36:587

Appleton's Cyc 6:454

Wheeler, Thomas M. (? -1895). Civil engineer and surveyor.

Eng Rec (1894-5) 31:399
RR Gaz (1895) 27:286

Whipple, Squire (1804-88). Civil engineer; surveyor for railroads and canals; constructed and wrote authoritative treatise on iron bridges.

ASCE Proc (1896) 22:558-61
ASCE Trans (1896) 36:527-30
Eng News (1888) 19:231*
Sci Am (1888) 58:224

Appleton's Cyc 6:462
DAB 20:70-1

Whistler, George Washington (1800-49). Railroad engineer, U.S. and Russia.

Amer Eng & RR J (1849) 22:313,582; (1852) 25:290-1
Assoc Eng Soc J (1886-7) 6:37-52

Appleton's Cyc 6:463*
DAB 20:72-3

Whitaker, Thomas Drake (1859-96). Mechanical engineer.

Eng & Min J (1896) 61:259

White, A. C. (1858-94). Electrical engineer; inventor of long distance transmitter.

Elec Eng (1894) 17:37

White, John Gardiner (? -1896). Civil engineer.

Eng & Min J (1896) 62:251
Eng Rec (1896) 34:271

White, Moores Merrick (1809-90). Pioneer builder of iron bridges.

Am Eng & RR J (1891) 65:43
Eng Rec (1890-1) 23:22
RR Gaz (1890) 22:865

White, William Howard (1847?-95). Civil engineer.

Eng Rec (1895-6) 33:39
RR Gaz (1895) 27:864

Whitehead, James R. (? -1892). Manufacturer of mechanical rubber goods.

Am Manuf (1892) 50:721

Whitehill, Robert (1845-93). Mechanical engineer, foundry and machine trades; later assistant engineer, U.S. Navy.

ASME Trans (1893) 14:1448
Eng Rec (1893) 28:118

Whitelaw, John (1831-92). Superintendent, Cleveland water works; city civil engineer.

ASCE Proc (1892) 18:191-2
Assoc Eng Soc J (1892) 11:538-9
Eng Rec (1892) 26:54

Whitman, Thomas Jefferson (1833-90). Civil and mechanical engineer, sewer and harbor work.

Am Eng & RR J (1891) 65:43
ASCE Proc (1892) 18:103-4
Assoc Eng Soc J (1891) 10:17-8
Eng Rec (1890) 22:406
RR Gaz (1890) 22:865

Whitmore, Frederick Culver (? -1898). Railway engineer for General Electric.

Elec Eng (1898) 26:654
Elec World (1898) 32:734

Whitney, Asa (1791-1874). Inventor of improvements in and manufacturer of cast-iron car wheels.

Iron Age (6/11/1874) 13:1
RR Gaz (1874) 6:237

Appleton's Cyc 6:487-8
DAB 20:155-6

Whitney, Eli (1765-1825). Inventor of cotton gin; developed method of firearms manufacture using principle of interchangeable parts.

Am Eng & RR J (1833) 2:454-8
Cassier's (1899-1900) 17:166-70*
Ind World (5/16/1895) 44:5-6
Mechanic's Mag (1833) 1:314-323

Appleton's Cyc 6:488-9*
DAB 20:157-60

Whitney, Samuel (? -1900). Engineer, river and harbor improvements in Wisconsin and neighboring states.

Eng Rec (1900) 41:601

Whiton, Augustus Sherrill (1821?-98). Civil engineer, railroads and iron rails.

Eng Rec (1897-8) 37:227
Iron Age (2/10/1898) 61:22
RR Gaz (1898) 30:109

Whittier, Charles (1829-99). Manufacturer of machinery.

ASME Trans (1900) 21:1158-9

Whittle, Louis Neale (1819?-86). Civil engineer; railroad director.

RR Gaz (1886) 18:152

Whitton, Andrew Dempster (1856-92). Civil engineer; specialized in construction of cable railways.

ASCE Proc (1894) 20:85-6

Whitwell, William Scollay (1809-99). Railroad engineer.

Eng Rec (1899) 40:563

Wiard, Norman (1820?-96). Inventor of and expert on ordnance and projectiles.

Eng & Min J (1896) 62:275
Eng Rec (1896) 34:289
Iron Age (1896) 58:542

Wickersham, J. B. (? -1892). Manufacturer of ornamental railings, Philadelphia.

Eng Rec (1891-2) 25:290

Wierman, Thomas T. (1834?-87). Civil engineer, railroads and canals in Pa.

Am Eng & RR J (1887) 61:398
Sanit Eng (1887) 16:270

Wiestling, George B. (1835-91). Official of Montalto Iron Company; civil engineer, Pennsylvania Railroad.

Am Eng & RR J (1891) 65:381
AISA Bull (1891) 25:187
Iron Age (1891) 47:1234
RR Gaz (1891) 23:438

Wightman, Henry M. (1840-85). City engineer, Boston.

ASCE Proc (1886) 12:124-5
Sanit Eng (1884-5) 11:399

Wightman, Willard Humphrey (1852-90). Civil engineer, Northern Pacific Railroad and Union Pacific Railroad.

ASCE Proc (1890) 16:217-8

Wilbraham, Thomas (1827-92). Engineer, railroads and steamboats.

ASME Trans (1892) 13:678

Wilcox, Stephen (1830-93). Inventor of safety water-tube boilers and of practical caloric or hot-air engine.

Am Mach (12/7/1893) 16:8
Am Manuf (1893) 53:942
ASME Trans (1894) 15:1188
Elec Eng (1893) 16:497
Eng (1893) 26:138
Iron Age (1893) 52:987,1176*
Power (12/1893) 13:15
RR Gaz (1893) 25:878
Sci Am (1893) 69:419

DAB 20:204-5

Wilkeson, John (1806-94). Inventor of stoves and heaters.

AISA Bull (1894) 28:75
Eng & Min J (1894) 57:324

Wilkinson, Alfred (1831?-86). Civil engineer, railroads.

RR Gaz (1886) 18:499

Wilkinson, Jephtha A. (1791?-1874). Inventor.

Sci Am (1874) 30:33

Wilkinson, John (1842-1900). Chief engineer,
several major buildings in N.Y. City.

ASME Trans (1900) 21:1163-4

Williams, Charles Penrose (? -1897). Mining
engineer and chemist.

Eng Rec (1897) 36:289

Williams, Edward Higginson (1824-99). Civil engineer,
railroads; employed by Baldwin locomotive works.

AISA Bull (1900) 34:4
Eng Rec (1899) 40:734
RR Gaz (1899) 31:900*

Williams, Isaac E. (1816?-83). Machinist and
locomotive engineer.

RR Gaz (1883) 15:733

Williams, Jesse Lynch (1807-86). Civil engineer,
canals in Ohio and Indiana.

AISA Bull (1886) 20:291
RR Gaz (1886) 18:711
Van Nostrand's (1871) 4:1-4*

Appleton's Cyc 6:523
DAB 20:268-9

Williams, John A. (1856?-98). Marine engineer.

Am Mach (1898) 21:415
Eng & Min J (1898) 65:648
Iron Age (5/26/1898) 61:15

Williams, Lloyd A. (1832?-74). Naval engineer.

Sci Am (1874) 30:33

Williams, Norman A. (1837-79). Civil engineer,
waterworks.

ASCE Proc (1881) 8:92

Williams, W. E. (1846?-94). Engineer; involved
in construction of rolling mills; supervised
mining operations, Mexico.

AISA Bull (1894) 28:285

Williamson, William Garnett (1840-98). Civil
    engineer, railroads in South and Mexico; city
    engineer, Montgomery, Ala.

    ASCE Proc (1899) 25:328-9
    ASCE Trans (1899) 41:645-6

Wilmarth, Seth (1810-86). Machinist and inventor of
    hydraulic lift for revolving turrets.

    Am Manuf (12/3/1886) 39:11
    Eng (1886) 12:135
    Eng & Min J (1886) 42:351
    Iron Age (11/11/1886) 38:15
    Sanit Eng (1886) 14:573

    Appleton's Cyc 6:543

Wilmot, Samuel R. (? -1897). Manufacturer of
    cold-rolled steel tubing and tools.

    Elec Eng (1897) 23:170
    Eng Rec (1896-7) 35:223

Wilson, Cary A. (? -1895). Chief engineer,
    Missouri, Kansas and Texas Railroad.

    RR Gaz (1895) 27:646

Wilson, John Allston (1837-96). Civil engineer,
    railroads and railroad bridges.

    ASCE Proc (1896) 22:150-1
    ASCE Trans (1896) 36:588-9
    Eng & Min J (1896) 61:91
    Eng Rec (1895-6) 33:129
    RR Gaz (1896) 28:65
    Sci Am (1896) 74:67

    Appleton's Cyc 6:553-4

Wilson, Theodore Delevan (1840-96). Naval
    contractor and shipbuilder.

    Am Mach (1896) 19:667
    Eng & Min J (1896) 62:11
    Eng Rec (1896) 34:81
    RR Gaz (1896) 28:478

    Appleton's Cyc 6:556
    DAB 20:346-7

Wilson, William (1832-91). Master mechanic and
    railroad engineer.

    Am Mach (9/10/1891) 14:5
    Nat Car (1891) 22:140

Wilson, William Willard (1841-87). Civil engineer and surveyor, railroads in Mexico and New York, waterworks in Yonkers.

Am Eng & RR J (1887) 61:103
ASCE Proc (1887) 13:40-1
Sanit Eng (1886-7) 15:267

Winans, Ross (1796-1877). Locomotive engineer; inventor of friction wheel, eight-wheel car and other improvements of railroad machinery.

Am Manuf (4/19/1877) 20:5
Sci Am (1877) 36:265

Appleton's Cyc 6:559
DAB 20:371-2
NYT (4/12/1877) 1:6

Winder, John C. (1832?-96). Civil engineer; worked on Croton Aqueduct, railways in South.

Eng & Min J (1896) 61:307
Eng Rec (1895-6) 33:291
RR Gaz (1896) 28:224

Winship, Ebenezer (1800?-67). Naval engineer and mechanic.

Sci Am (1867) 17:407

Winslow, Ephraim N. (1824-80). Chief engineer, Old Colony Railroad Company, Mass.

RR Gaz (1881) 13:35

Winslow, John Flack (1810-92). Iron manufacturer, Troy, N.Y.; builder of Monitor.

AISA Bull (1892) 26:68
Eng Rec (1891-2) 25:242
Iron Age (1892) 49:513
RR Gaz (1892) 24:218

DAB 20:399-400

Winter, Herman (1829-95). Marine steam engineer; shipbuilder and designer.

ASME Trans (1896) 17:740-1
Eng (1895) 30:66
Eng Rec (1895) 32:273
Iron Age (1895) 56:543

Witt, Stillman (1808-75). Engineer, bridges and railroads.

RR Gaz (1875) 7:199

NYT (5/4/1875) 6:7

Wolcott, Alexander (1814-84). Civil engineer and surveyor, Chicago.

Assoc Eng Soc J (1884-5) 4:32-3

Wolfe, Abraham N. (? -1895). Inventor and manufacturer of turbine water wheels and milling machinery.

Eng Rec (1894-5) 31:273

Wollensak, J. F. (1847-90). Inventor and manufacturer of hardware.

Iron Age (1890) 46:113*

Wood, Albert D. (? -1895). Civil engineer, employed by railroads and oil producing and refining companies.

Eng Rec (1895-6) 33:75

Wood, Allen N. (? -1892). Manufacturer of machinery.

Iron Age (1892) 50:583

Wood, Arthur Hastings (1869-92). Civil engineer.

ASCE Proc (1892) 18:106

Wood, Charles (1862-95). Civil engineer, Cincinnati, Hamilton & Dayton Railroads.

ASCE Proc (1896) 22:697-9
ASCE Trans (1896) 36:591-3
Eng Rec (1895-6) 33:2
RR Gaz (1895) 27:812

Wood, De Volson (1832-97). Professor of mechanical and civil engineering at several institutions; author of texts.

Am Eng & RR J (1890) 19:79*
ASME Trans (1897) 18:1106-9
Elec Eng (7/8/1897) 24:20
Eng & Min J (1897) 64:12
Eng News (1897) 38:14*
Eng Rec (1897) 36:91
Iron Age (7/1/1897) 60:16
RR Gaz (1897) 29:477
Sci Am (1897) 77:19

Appleton's Cyc 6:591-2

Wood, John (1816-98). Iron manufacturer; first
maker of sheet iron in U.S.

AISA Bull (1898) 32:93
Eng & Min J (1898) 65:678
Iron Age (6/2/1898) 61:32

Wood, Joseph (1819?-99). Locomotive engineer.

AISA Bull (1899) 33:188
Am Mach (1899) 22:1026
RR Gaz (1899) 31:736

Wood, Joseph H. (1838-1900). Engineer, canals
and railroads.

RR Gaz (1900) 32:643

Wood, S. A. (? -1900). Mechanical engineer.

Am Manuf (1900) 66:307

Wood, Thomas (1814?-99). Founder of Fairmount
Machine Works, Philadelphia.

Am Mach (1899) 22:419

Wood, Walter Abbott (1815-92). Inventor and
manufacturer of harvesting machinery.

Am Mach (1/28/1892) 15:8
Iron Age (1892) 49:117
Sci Am (1892) 66:97

Appleton's Cyc 6:597-8
DAB 20:475-6

Wood, W. Dewees (1826?-98). Manufacturer of sheet
iron, Pittsburgh.

Am Eng & RR J (1899) 73:64
AIME Trans (1900) 30:xxxix-xl
AISA Bull (1899) 33:12
Am Mach (1899) 22:37
Eng & Min J (1899) 67:61
Eng Rec (1899) 39:128
Iron Age (1/12/1899) 63:14

Wood, William Maxwell (1850-97). Inventor of mechanical
ice-making machinery and boat-detaching apparatus.

Elec Eng (1897) 24:636
Iron Age (12/23/1897) 60:21

Appleton's Cyc 6:599

Woodling, W. H. (1821?-86). Manufacturer of freight cars.

RR Gaz (1886) 18:782

Woodruff, Samuel (1814?-82). Manufacturer of mining and other heavy machinery.

RR Gaz (1882) 14:200

Woodruff, Theodore (1811?-92). Reputed inventor of sleeping car.

Iron Age (1892) 49:930

Woods, Arthur T. (1859-93). Mechanical engineer, railroads and marine works; professor; associate editor of Railroad Gazette.

ASME Trans (1893) 14:1443-4
Eng & Min J (1893) 55:132
Eng Rec (1892-3) 27:211
Iron Age (1893) 51:373
Nat Car (1893) 24:40
RR Gaz (1893) 25:112

Woodward, Ezekiel W. (? -1898). Civil engineer, railroads.

RR Gaz (1898) 30:570

Woodward, Jabez Marshall (1808?-91). Civil engineer, railroads in Illinois.

Eng Rec (1891-2) 25:2

Wootton, Edwin H. (1835?-92). Developer of asphalt for engineering and construction purposes.

Eng Rec (1892) 26:106

Wooten, John E. (1822-98). Mechanical engineer; inventor of firebox and other engineering devices.

Am Eng & RR J (1899) 73:29
Am Mach (1898) 21:965
Eng & Min J (1898) 66:764
Iron Age (12/22/1898) 62:17
RR Gaz (1898) 30:922,927

Worcester, Franklin E. (1860-91). Mechanical engineer.

ASME Trans (1891) 12:1056-7

Worden, Charles A. (? -1898). Military engineer.

Eng Rec (1898) 38:377

Worrall, James (1812?-85). Engineer, railroads
and canals.

RR Gaz (1885) 17:254

Worth, Gustavus A. (? -1897). Railroad engineer.

RR Gaz (1897) 29:735

Worthen, William Ezra (1819-97). Civil engineer,
hydraulic work; expert on pumping machinery
and on flow of water in canals.

ASCE Proc (1898) 24:709-11
ASCE Trans (1898) 40:565-7
Eng & Min J (1897) 63:359
Eng News (1887) 17:80*
Eng Rec (1896-7) 35:399
Iron Age (4/8/1897) 59:18
RR Gaz (1897) 29:259

Appleton's Cyc 6:617
DAB 20:538-9

Worthen, William H. (1842-92). Mechanical engineer;
inventor of pumping machinery and condensing
apparatus.

ASME Trans (1892) 13:682

Worthington, George (1844-92). Electrical engineer.

Am Mach (2/25/1392) 15:8
Elec Eng (1892) 13:146

Worthington, Henry Rossiter (1817-80). Mechanical
and hydraulic engineer; worked on canal navigation
and city water supply, N.Y. City; developed steam
pumps and boilers.

ASME Trans (1881) 2:234
Eng (1881) 1:5
Eng & Min J (1880) 30:17*,409-10
Eng News (1880) 7:441; (1893) 29:287*
Iron Age (1880) 26:14
RR Gaz (1880) 12:690
Sanit Eng (1880-1) 4:85
Sci Am (1881) 44:3

Appleton's Cyc 6:617
DAB 20:539

Wright, Franklin (? -1886). Civil engineer and surveyor, railroads.

Am Manuf (2/26/1886) 38:12
RR Gaz (1886) 18:152

Wright, George F. (1848-92). Civil engineer, railroads and hydraulic works.

ASCE Proc (1892) 18:206-7

Wright, Horatio Gouverneur (1820-99). Military engineer; responsible for East River Bridge, Sutro Tunnel (Nevada), Washington Monument.

Eng Rec (1899) 40:135
RR Gaz (1899) 31:498

DAB 20:554-5

Wright, John A. (1820?-91). Railroad engineer.

Eng Rec (1891) 24:360

Wright, Lysander (1822?-98). Manufacturer of engines and machinery.

Am Mach (1898) 21:867
Iron Age (11/17/1898) 62:22

Wright, Thomas W. (? -1897). Civil engineer and surveyor.

Eng Rec (1897) 36:553

Wright, William E. (? -1891). Marine engineer.

Eng (1891) 22:5

Wright, William Wierman (1824-82). Military and civil engineer, bridges, railroads and roads.

ASCE Proc (1882) 8:119-20
RR Gaz (1882) 14:171

Yale, Charles Oscar (1831?-95). Inventor of safe and lock vaults.

Iron Age (1895) 56:899

Yates, John B. (? -1899). Civil and military engineer, railroads and bridges.

Eng Rec (1899) 40:514

Yeatman, Henry Clay (1866-96). Engineer, railroads and waterworks.

ASCE Proc (1898) 24:781-2
ASCE Trans (1898) 40:579

Yerkes, E. A. (? -1900). Improver of machine manufacture of hammers, sledges and edge tools.

Am Manuf (1900) 67:68

Yorke, Edward (1835-84). Civil engineer, railroads in Mexico and West.

ASCE Proc (1885) 10:124-6

Yorke, William (? -1897). Gas engineer.

Eng Rec (1896-7) 35:531

Young, Herbert Andrew (1857-94). Civil engineer, railroads in New York and Mexico.

ASCE Proc (1894) 20:204-5; (1895) 21:69

Young, Robert Alexander (? -1893). Civil engineer.

Eng Rec (1893) 28:246

Young, William Clark (1799-1893). Chief engineer, Hudson River Railroad, etc.; pioneer in railroad construction and operation.

ASCE Proc (1892) 19:192-4
Eng News (1893) 30:514
Eng Rec (1893-4) 29:68
RR Gaz (1893) 25:957

Younger, James (1844?-94). Consulting engineer, Cramp Shipbuilding Company.

Eng & Min J (1894) 57:420
Eng Rec (1893-4) 29:360
Iron Age (1894) 53:851
RR Gaz (1894) 26:328

Yule, George (1842?-1900). Inventor and manufacturer of hat-making machinery.

Am Mach (1900) 23:305

Ziegler, J. Q. A. (? -1885). Chief engineer, U.S. Navy.

Eng (1885) 10:19

Zimmerman, G. F. S. (1826-96).   Inventor; manu-
    facturer of iron castings and threshing machines.

    Iron Age (1896) 58:1265